T0321017

Bio-Informatic Systems, Processing and Applications

RIVER PUBLISHERS SERIES IN COMMUNICATIONS

Consulting Series Editors

MARINA RUGGIERI
University of Roma "Tor Vergata"
Italy

HOMAYOUN NIKOOKAR
Delft University of Technology
The Netherlands

This series focuses on communications science and technology. This includes the theory and use of systems involving all terminals, computers, and information processors; wired and wireless networks; and network layouts, procontentsols, architectures, and implementations.

Furthermore, developments toward newmarket demands in systems, products, and technologies such as personal communications services, multimedia systems, enterprise networks, and optical communications systems.

- Wireless Communications
- Networks
- Security
- Antennas & Propagation
- Microwaves
- Software Defined Radio

For a list of other books in this series, visit
http://riverpublishers.com/river publisher/series.php?msg=Communications

Bio-Informatic Systems, Processing and Applications

Editors

Johnson I Agbinya
Edhem Custovic
Jim Whittington
Sara Lal

River Publishers

Aalborg

ISBN 978-87-93102-18-7 (hardback)

Published, sold and distributed by:
River Publishers
Niels Jernes Vej 10
9220 Aalborg Ø
Denmark

Tel.: +45369953197
www.riverpublishers.com

This book is dedicated to all the professors and teachers who have had input in our professional training and development.

Table of Contents

Preface

Engineering systems are often a combination of several technologies brought together to solve a problem in society. This book is a summary of state-of-the-art biomedical engineering and computing technologies which are desirable for implementing imaginative new biomedical systems. More and more medical procedures and diagnosis which used to be exclusively undertaken by human beings are either being done completely by machines or with the assistance of electronics and software systems. The rationale for this paradigm shift is motivated by the ability of smart computing and electronic machines to see through a maze of options and better data analysis compared with a human being. Hence reliance on bio-medical data processing and analysis will become more and more ubiquitous as we progress into the future.

The book is a summary of state-of-the art and emerging biomedical algorithms and system which are nearing maturity. Enhanced cancer detection through the application of neural networks for 'prediction of cancer based on tumor markers values' is presented in chapter 1. In chapter 2 the authors present an advanced method for control of upper-limb prostheses using myoelectric signals (MES), which are directly obtained from the skin-surface. The measured MES is filtered and pre-processed before pattern recognition and classification methods are applied, to uncover prehensile motions or gestures. Biomedical research has so far focused on what happens in hospitals and laboratories. Hence software and hardware systems often target such environments. However smart home will become a feature of emerging home health care delivery. Therefore developing cheap, reliable and resilient hardware and software combinations that can be deployed to the field and in particular the home environment is essential. That is the subject of chapter 3. Chapter 4 adds to the progress in chapter 3 in terms of how to detect falls that could arise in home-based health care delivery. The combination of sensor networks and the Internet will support remote home-based health care and bio-signal monitoring and telemetry. Chapter 5 describes such a system with the main goal of developing a"Smart Hospital" and define its main performance requirements using ZigBee sensor network deployed in a real hospital environment. The processing and applications of electromyography and electroencephalography in fatigue detection to support safe transportation are covered in chapters 6 to 8. By detecting the state of a vehicle driver, such as drowsiness the authors have shown that conditions which result to accidents can be minimized to increase the safety of road users. Chapters 9, 10 and 11 are applications of biomedical engineering in medical training and support.

In addition to the high level research findings in these chapters, the fundamental principles covered in the book lend themselves to easy application in other medical fields. Undergraduate and postgraduate students in bio-informatics, biomedical electronics, health practitioners and biomedical system developers will benefit tremendously by applying the techniques and results presented.

Acknowledgment

This book is made possible by the experts who reviewed conference papers for the international conference on broadband communications and biomedical applications (IB2COM) between 2009 and 2011. More specifically we acknowledge the contributions of the host institutions over the three years including La Trobe University, Melbourne Australia, the University of Technology, Wroclaw Poland, the University of Malaga, Malaga Spain and the University of Technology, Sydney Australia. We acknowledge the use of the cover art which was designed and produced by Jerry Ijale Agbinya. We are grateful for his contribution.

.

Chapter 1.

Neural Networks Based System for Cancer Diagnosis Support

Witold Jacak, Karin Proell

Upper Austrian University of Applied Sciences, Faculty of Informatics, Communications and Media, 4232 Hagenberg, Austria
witold.jacak@fh-hagenberg.at , karin.proell@fh-hagenberg.at

1.1   INTRODUCTION

Tumor markers are substances produced by cells of the body in response to cancerous but also to noncancerous conditions. They can be found in body liquids like blood or in tissues and can be used for detection, diagnosis and treatment of some types of cancer. For different types of cancer different tumor markers can show abnormal values and the levels of the same tumor marker can be altered in more than one type of cancer. Examples of tumor markers include CA 125 (in ovarian cancer), CA 153 (in breast cancer), CEA (in ovarian, lung, breast, pancreas, and gastrointestinal tract cancers), and PSA (in prostate cancer). Although an abnormal tumor marker level may suggest cancer, tumor markers are not sensitive or specific enough for a reliable cancer diagnosis. But abnormally altered tumor marker values indicate a need for further medical examination.

During blood examination only a few tumor marker values are tested and for this reason the usage of such incomplete data for cancer diagnosis support needs estimation of missing marker values. Neural networks are proven tools for prediction tasks on medical data [1,2,7,9]. For example neural networks were applied to differentiate benign from malignant breast conditions based on blood parameters [1], for diagnosis of different types of liver disease [7], for early detection of prostate cancer [2,8], for studies on blood plasma [7] or for prediction of acute coronary syndromes [3]. In this chapter we present two

Bio-Informatic Systems, Processing and Applications , 1-26,

approaches for a system to support cancer diagnosis. Both systems use heterogeneous neural networks, the first one uses tumor marker values and tumor diagnosis data of thousands of patients for training and testing the neural networks for cancer prediction [6] and the second one extends the tumor marker values by standard blood parameters. Unfortunately neural networks are unable to work properly with incomplete data; missing values however are a common problem in medical datasets. It may be that a specific medical procedure was not considered necessary in a particular case or that the procedure was taken in a different laboratory with the values not available in the patient record, or that the measurement was taken but not recorded due to time constraints. A main focus in this work is laid on this problem of missing values in biomedical data as they make training of neural networks difficult.

The cancer prediction system is based on data coming from vectors $C = (C_1, ..., C_n)$ containing tumor marker values which are frequently incomplete, containing lots of missing values influencing the plausibility of diagnosis prediction. The question rises if it is possible to increase the quality of cancer prediction by using information beyond tumor marker data. The general goal of a data driven cancer prediction system can be expressed as follows:

Construct a data driven cancer diagnosis support system which:

- maximizes the probability of correct cancer diagnosis (positive and negative)
- minimizes the probability of incorrect diagnosis if a cancerous disease exists

One efficient method for such a system is the synthesis of complex neural networks for prediction of cancer based on tumor markers values. We need many thousand datasets for training and evaluating the neural networks. As mentioned before these datasets contain lots of missing values. To overcome this problem we additionally make use of datasets containing a whole blood parameter vector $P = (P_1, ..., P_m)$ of each patient. Frequently those vectors are incomplete too. For these reasons we link two independently trained neural network systems into one: The first subsystem is trained only with complete or incomplete tumor marker datasets $C = (C_1, ..., C_n)$. The second one includes also blood parameter vectors $P = (P_1, ..., P_m)$ to support prediction of cancer possibility.

1.2 TUMOR MARKER VALUES BASED CANCER DIAGNOSIS SUPPORT SYSTEM

Cancer diagnosis support uses parallel working systems ($Cancer_k$), with the same structure of networks trained for different types of cancer. The input of each $Cancer_k$ system is the complete or incomplete vector C of tumor marker specific for the chosen type of cancer, and the output represents the possibility (values between 0 and 1) of a cancerous disease. Output values of the network system greater than 0,5 are treated as cancer occurrence.

Each *Cancer$_k$* system consists of many different groups of neural networks (see Figure 1.1).

- Group of neural networks (C_{net}) for individual marker C_i: $i=1,..,n$.
- Feed forward neural network ($C_{Group}FF_{net}$) for vector of marker C, with complete or incomplete values.
- Pattern recognition neural network ($C_{Group}PR_{net}$) for vector of marker C, with complete or incomplete values.
- Cascaded coupled aggregation method for final calculation of cancer plausibility.

1.2.1 Group of separate neural networks for individual marker (C_{net})

The first group of neural networks contains parallel coupled neural networks, which are individually trained for different tumor markers. Each neural network

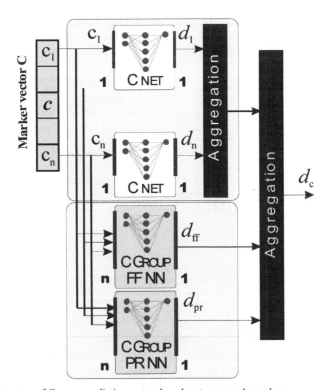

Figure 1.1 Structure of *Cancer$_k$* prediction system based on tumor marker values.

is of type feed forward with one hidden layer having 6 -10 neurons, activation functions tan/sigmoid, further one input (normalized tumor marker value) and one output (diagnosis: 0 – no cancer (healthy) and 1– cancer (ill)).

The networks were trained independently of type of cancer disease (i.e. for all types of cancer diseases).

The values of markers are further categorized into four intervals (Classes). The first interval includes all values less than a *Normal Value* of marker, the second interval includes all values between the *Normal Value* and an *Extreme Normal Value* of marker, third interval includes values between the *Extreme Normal Value* and a still *Plausible Value* of marker and fourth interval includes all values greater than *Plausible Value*.

The input values of each network for each training and testing process are normalized using the respective upper bound of *Plausible Value*. Each value of marker, which extends that upper bound, obtains the value 1. The individually trained networks represent a generalized cancer occurrence prediction, disregarding specific type of cancer and based only on one specific tumor marker. Examples of trained neural networks (with 6 neurons in hidden layer) for tumor markers C125, C153, C199 and CEA are presented in Figure 1.2. The x-axis represents the normalized values of tumor markers and the y-axis represents cancer possibility. Figure 1.2 presents the outputs of the trained networks C125 net, C153 net, C199 net and CEA net, respectively. Network-output value greater as 0,5 (blue line) is interpreted as cancer occurrence.

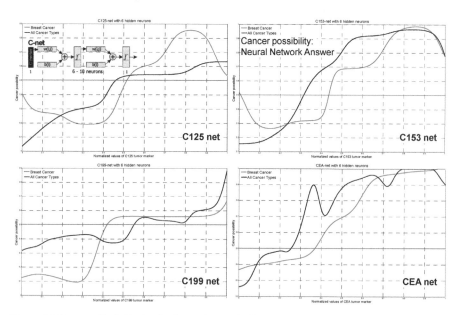

Figure 1.2 Outputs of individual trained neural networks for tumor markers C125, C153, C199 and CEA.

The networks were trained with 2598 datasets for C125 marker, with 2442 datasets for C153 marker, with 4519 dataset for C199 marker and with 7153 dataset for CEA marker (black curve). The datasets contain data with different cancer types from C00 to C96 ICD 10 code (44% of all datasets) and data without cancer occurrence (56% of all datasets).

Additionally, we trained these networks with smaller sets of data for one specific type of cancer disease. Figure 1.2 presents examples of trained C125, C153, C199 and CEA markers networks representing breast cancer (red curve).

Generally the network trained for all cancer types is more pessimistic, it means this network predicts cancer possibility greater than 0,5 for smaller values of tumor marker as the network trained for one special cancer type. In the example in Figure 1.2 the threshold points of networks trained for all cancer types is 0,46 (75,9 U/ml) for C125 marker. For marker C153 the threshold was 0,32 (34,8 U/ml), for C199 0,55 (73,1 U/ml) and or CEA 0,25 (14,1 ng/ml). The threshold points of networks trained for breast cancer are 0,52 (85,5 U/ml) for C125, 0,44 (48,2 U/ml) for C153, 0,42 (55,8 U/ml) for C199 and 0,37 (20,4 ng/ml) for CEA. The regressions between network outputs trained for all cancer types and breast cancer type are 0.69, 0.88, 0.88, and 0.91 for C125, C153, C199 and CEA, respectively. The influence of networks trained in this way on the final diagnosis prediction will be discussed in the next section.

The input of parallel coupled C_{nets} is the vector of tumor marker $\mathbf{C} = (C_1, ..., C_m)$, where some C_i are missing. When the tumor marker value in vector $\mathbf{C}$ is available, then the adequate C_{net} calculates the predicted cancer possibility. When a marker value in vector $\mathbf{C}$ is not available, then the output of C_{net} is set to -1. The individually calculated output values of C_{nets} can be aggregated in many different ways. We compare three methods of aggregation:

- Maximum value of all individual network outputs:

$$C_{net}(\mathbf{C}) = \max\{ C^i_{net}(C_i) | i = 1, .., m\} \tag{1.1}$$

- Average value of all individual network outputs, without missing values:

$$C_{net}(\mathbf{C}) = \operatorname{avg}\{ C^i_{net}(C_i) | i = 1, .., m \ \& \ C_i \neq -1\} \tag{1.2}$$

- $net_{aggregation}$ - neural network trained on individual networks outputs (this neural network can be trained with data of only one chosen cancer type $Cancer^k$).

$$C_{net}(\mathbf{C}) = net_{aggregation}(C^i_{net} | i = 1, .., m) \tag{1.3}$$

Other interesting possibility is using the thresholded values of outputs of individual networks (i.e. if $C_{net}(C) < 0,5$ then $C_{net}(C) = 0$, else $C_{net}(C) = 1$) as input for a perceptron type network.

We use one aggregation type in the full system. In case of *max* aggregation: If only one marker of the marker group shows a greater value than the aggregation has yielded this value is taken.

The diagnosis prediction based on aggregation of separate cancer predictions of individual marker networks C_{net} is not sufficient for the generalization of cancer occurrence. It is necessary to reinforce the information coming from data of the whole group of markers. Therefore two neural networks with cumulative marker groups are added. These networks will be trained only for a specific cancer type.

1.2.2   Feed forward and pattern recognition neural networks for tumor marker group

The vectors C of marker values can again be incomplete. If a tumor marker value in vector C is missing, then this value is set to -1. This allows to generate training sets for a specified cancer type *Cancer$_k$* and to train the two neural networks (Figure 1.1):

- Feed forward neural network with 16-20 hidden neurons and tansig/linear activation functions ($C_{group}FF_{net}$)
- Pattern recognition network with 16-20 hidden neurons ($C_{group}PR_{net}$)

The outputs of all parallel working networks are coupled into a new vector and this represents an input for the cascade-net ($Cascade_{net}^{k-Cancer}$) for diagnosis generalization with 16 hidden neurons or other aggregation function such as *mean* or *max*.

1.3 CASE STUDY: TUMOR MARKERS BASED BREAST CANCER DIAGNOSIS SUPPORT SYSTEM

1.3.1   Setup and Results

Training and test datasets were prepared for breast cancer. For tumor marker group we have taken the C125, C153, C199 and CEA markers ($C = (C_{125}, C_{153}, C_{199}, C_{EA})$). The training and test datasets include about 5100 and 2480 data, respectively. Target outputs have binary values, 0- no cancer and 1 – cancer. For individual C^i_{nets} we use networks trained for all types of cancer. The outputs of individual networks are aggregated with the *max* function and with a separately trained *perceptron* network for breast cancer. $C_{group}FF_{net}$ and $C_{group}PR_{net}$ networks are trained only for breast cancer. The confusion matrices between target diagnosis and outputs of networks used are shown in Figure 1.3.

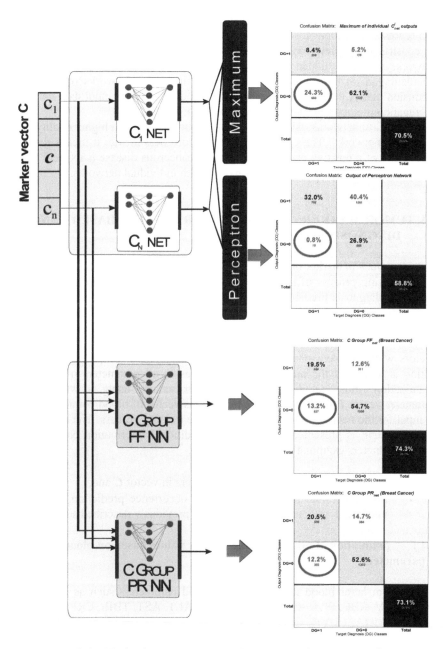

Figure 1.3 Confusion Matrices between target diagnosis and outputs of separate networks.

The outputs of all parallel working networks are coupled into a new vector and this represents an input for the diagnosis generalization system. This system can be constructed as new cascade-net (Cascade$_{net}^{k\text{-Cancer}}$) pattern recognition type network, with 16 hidden neurons or with a classical aggregation function calculating *mean* or *max* values of coupled first level networks outputs. The confusion matrix between test target data and outputs on test input data for these aggregation methods are presented in Figure 1.4.

All neural networks using aggregation predict cancer with higher quality than individual networks. The coupled system is more pessimistic, it means that the probability of a positive prediction in case no cancerous disease is existent (false positives) is greater than the prediction done with individual networks.

1.4 TUMOR MARKERS AND BLOOD PARAMETERS BASED CANCER DIAGNOSIS SUPPORT SYSTEM

Missing values in marker data make datasets incomplete, which leads to problems in the training process of neural networks and in consequence to a decrease of quality of diagnosis prediction. Many of the available marker vectors consist of just a few measurements and cannot be used as training data for neural networks without further processing [4,5].

One approach to overcome this problem is to restrict the analysis only to vectors with complete data but this leads to very small sample sets. Another option is to extend the number of input values to all parameters of the blood examination, thus including also non-marker values, using a whole blood parameter vector $P = (P_1, ..., P_m)$ as input. This vector of blood parameters used as input for the neural networks can support the training process. The structure of such a system is presented in Figure 1.5. Additional information coming from blood parameters examination P can be used to:

- estimate missing value of tumor markers in vector C and,
- train additional networks for cancer occurrence prediction, which are integrated into one system containing previously described subsystems

1.4.1 Estimation of missing tumor marker values based on blood parameters

Typically in labor blood examination 27 blood parameters such as HB, WBC, HKT, MCV, RBC, PLT, KREA, BUN, GT37, ALT, AST, TBIL, CRP, LD37, HS, CNEA, CMOA, CLYA, CEOA, CBAA, CHOL, HDL, CH37, FER, FE, BSG1, TF and tumor markers such as AFP, C125, C153, C199, C724, CEA, CYFRA, NSE, PSA, S100, SCC, TPS etc. are measured. For each parameter and marker there are experimentally established upper and lower bounds of values (see Table 1.1)

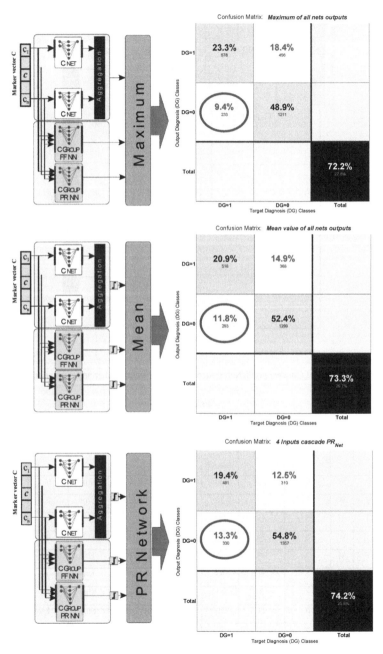

Figure 1.4 Confusion Matrices between target diagnosis and aggregated outputs of parallel working networks.

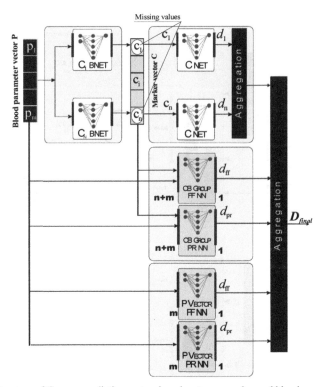

Figure 1.5 Structure of $Cancer_k$ prediction system based on tumor marker and blood parameters data.

Table 1.1
Example of value boundaries of blood parameters

CODE	Sex	Normal Lower Bound	Normal Upper Bound	Extreme Norm. Upper Bound	Over Norm. Plausible Upper Bound	Type	Unit	Age LB	Age UP
AFP	M	0	5,8	28	99	AFP (CL)	IU/ml	0	199
AFP	W	0	5,8	28	99	AFP (CL)	IU/ml	0	199
ALT	M	5	45	135	247,5	ALT (GPT)	U/l	0	199
ALT	W	5	34	102	187	ALT (GPT)	U/l	0	199
AST	M	5	35	105	192,5	AST (GOT)	U/l	0	199
AST	W	5	31	93	170,5	AST (GOT)	U/l	0	199
BSG1	M	3	8	15	55	Sinking 1h	Mm	1	199
BSG1	W	6	11	20	55	Sinking 1h	Mm	1	199
BUN	M	5	18	50	165	BUN	Mg/dl	2	16
BUN	W	5	18	50	165	BUN	Mg/dl	2	16

Table 1.2
C153 tumor marker values

Code	Class	μ	σ	Min	Max	Dmax
C153	1	15,54	5,02	2	25	100
C153	2	33,59	6,73	26	50	100
C153	3	68,48	13,78	56	100	100
C153	4	162,20	321,22	101	10000	100

We divide the values range of marker C and blood parameter P into k non-overlapping intervals, called classes.

In our case study we define four classes ($k = 4$). *Class 1* includes all values less than *Normal Value* of marker or blood parameter, *Class 2* includes all values between *Normal Value* and *Extreme Normal Value* of marker or blood parameter, *Class 3* includes values between *Extreme Normal Value* and *Plausible Value* of marker or blood parameter and *Class 4* include all values greater than *Plausible Value*.

For each class i of marker values we calculate the average value μ_i and standard deviation σ_i. For example the respective values of marker C153 calculated from patient data are presented in Table 1.2.

These classes and their boundaries are used for normalization of marker and blood parameter values: Input and output values of each network for training and testing process are normalized using the respective upper bound of *Plausible Value*. Each value of parameter or marker, which is greater than its upper bound, obtains the normalized value 1 ensuring that all values belong to interval [0, 1].

1.4.2   System for estimation of tumor marker values

The general marker estimation system consists of three neural networks (Figure 1.6):

- Feed forward neural network (FF) with p inputs (normalized values of blood parameter vectors P) and one output, normalized values of marker C_i
- Pattern recognition neural network (PR) with p inputs (normalized values of blood parameter vectors P) and k outputs, k-dimensional binary vector coding classes of marker C_i
- Combined feed forward neural network (FC) with p inputs (normalized values of blood parameter vectors P) and two outputs:
  - normalized values of marker C_i (as in network FF), and
  - normalized classes of marker C_i as:

$$\text{NormClass}^j(C_i) = j/k \qquad \text{for } j=1,...,k \qquad (1.4)$$

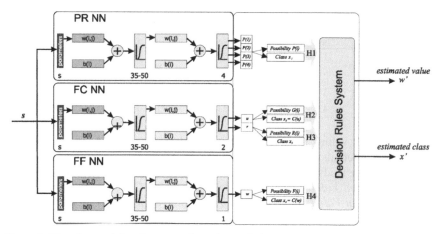

Figure 1.6 Neural network based tumor marker value prediction system.

All neural networks have one hidden layer and tan-sigmoid or log-sigmoid transfer function. The output values of neural networks belong usually to interval [0, 1].

- The pattern recognition neural network PR produces the k-value vector $(\pi(i) \mid i = 1,..,k)$, where $\pi(i)$ describes the possibility (in sense of fuzzy logic) of class i of C marker connected to input parameter vector $\boldsymbol{P}$. The class of marker C, which will be supposed is

$$x_1 = \arg(\max\{\pi(i) \mid i=1,...,k\}) \tag{1.5}$$

- The feed forward neural network FC uses value and class as inputs and generates two outputs: a value r interpreted as possibility of class of marker C and a value u representing the prediction of the normalized value of marker C. Value u and the known limits of the marker values are used for determination of class $x_2 = Class(u)$ indicating the class to which the output u belongs. Separately we calculate the possibility vector $(G(i) \mid i = 1,...,k)$ as indirect distance to the average value of each class of marker C

$$G(i) = 1-(|u-\mu_i|/d_{max}) \qquad \text{for } i = 1,...,k \tag{1.6}$$

where d_{max} denotes maximal distance between values of marker C.
For value r we calculate the possibility vector $(R(i) \mid i = 1,...,k)$ as

$$R(i) = 1-|r-i/k| \qquad \text{for } i = 1,...,k \tag{1.7}$$

The class of marker C that will be supposed is

$$x_3 = \arg(\max\{R(i)|i=1,...,k\}). \tag{1.8}$$

- The feed forward neural network generates a normalized value w of marker C. Based on value w and the limits of each class we can calculate the class $x_4 = Class(w)$ to which belongs the output of FF network.

  Separately we calculate the possibility vector $(F(i) \mid i = 1,...,k)$ as indirect distance to average value of each C marker class

$$F(i) = 1-(|w-\mu_i|/d_{max}) \qquad \text{for } i = 1,...,k. \tag{1.9}$$

1.4.3   Evaluation and post processing method

Based on the calculated estimation of marker values we can establish four hypotheses x_1, x_2, x_3, x_4 for determination of classes. For each hypothesis x_1, x_2, x_3, x_4 the possibility value π, G, R, F is calculated too.

These hypotheses should be verified to find the maximal possible prediction. This is done by testing a couple of aggregation functions V on the possibility values of each hypothesis. Those aggregation functions are similar to aggregation rules in fuzzy logic decision-making systems. In this experiment we use four kinds of functions V:

Minimum: $V(x_i) = \min\{\pi(x_i), G(x_i), R(x_i), F(x_i)\}$

Product: $V(x_i) = \pi(x_i) \cdot G(x_i) \cdot R(x_i) \cdot F(x_i)$

Average: $V(x_i) = (\pi(x_i)+ G(x_i) + R(x_i) + F(x_i))/4$

Count: $V(x_i) = Count\_of\,(x_i)$

where Count is the number of identical hypothesis.

The maximal value of aggregation functions determines the new predicted class. i.e.

$$x_{new} = \arg\,(\max\{\,V(x_i)\mid i=1,..,4\}) \tag{1.10}$$

where $V(x_i) = V(P(x_i), G(x_i), R(x_i), F(x_i))$ is the evaluation function of arguments $P(x_i), G(x_i), R(x_i)$ and $F(x_i)$. This kind of evaluation method leads to four decision composition rules, namely *MaxMin, MaxProd, MaxAvg* and *MaxCount* are well known in fuzzy decision systems. Based on different experiments it could be observed that *MaxAvg* and *MaxMin* performed best depending on dataset.

Estimation of predicted marker value: Based on the determined new class the estimation of a marker value is performed. If the evaluation function V used in decision composition rule has a value greater than τ then the new estimated value of a marker is equal to the average value of w and u. i.e.

$$w_{new} = (w + u)/2 \qquad \text{if } V(x) > \tau \tag{1.11}$$

$$w_{new} = (w + u + \mu_{xnew})/3 \qquad \text{other} \tag{1.12}$$

1.4.4   Case Study: C153 Tumor Marker

Training and test setup: We have taken the complete data set containing 20 blood parameters form patient data. The input vector **P** contains complete data of following blood parameters **P** = (*HB, WBC, HKT, MCV, RBC, PLT, KREA, BUN, GT37, ALT, AST, TBIL, CRP, LD37, HS, CNEA, CMOA, CLYA, CEOA, CBAA*). We used 4427 samples as Learning Pattern Set of neural networks system and 491 independent samples as Test Set. Based on whole data set the Pearson correlation coefficient between tumor markers and blood parameter is calculated. This correlation matrix in % is shown in Table 1.3.

Table 1.3
Correlation between tumor markers and blood parameters

Marker/ Parameter	AFP	C125	C153	C199	C724	CEA	CYFS	FPSA	NSE	PSA	PSAQ	S100	SCC	TPS
ALT	16	-10	25	19		10			31			-10		49
AST	33	13	40	27	15	27	11		49			8		36
BSG1			22	13	9	50	13	34		23	15	119	25	15
BUN		17					25	18	15		19	74	10	8
CBAA			-10									-11		
CEOA	-11		-13		-11									9
CH37	-28	-39		-23	-16	-9	-35	-11	-21				-12	-13
CHOL		-20			14		-13					-39		
CLYA	-19	-36	-20	-12	-20	-16	-15		-14	-13	-9	-35	-14	
CMOA		18			19	10								8
CNEA		25	19		11	12	17	10	11	11		21	14	
CRP	18	32	27	17	23	23	20		41	21			11	
FE	11	-18	-29		-24	-12	-16					-11		-27
FER	15	31	31	32			35		59		14			-11
GT37	22	15	37	39	50	30			43					30
HB		-35	-38	-25		-19	-18			-15		-12	-17	
HDL	-11			-16								-151		24
HKT		-32	-36	-25		-16	-13		-8	-14		-16		
HS	-15	19		-12	-15						9			9
KREA		15	8	-9	9		14	17			21	64	10	
LD37	19	34	45	23	29	35	40		69	14		70		
MCV	17			10								-12		
PLT	-15	10	16			12	16				-11	17	12	12
RBC		-30	-31	-28		-16	-9			-12		-11		
TBIL	18	13		19	-16	12			21		9			
TF		-55	-29	-26	-28	8	-32	-20	-25				-12	34
WBC							22	11	17			24	14	

Data normalization: All data are normalized (on values [0, 1]) based on 3 classes (*Normal Value, Extreme Normal Value* and *Over Normal but Plausible Value*). Values higher *Plausible Value* are the normalized to value 1. Data taken for learning and test process include all four classes.

1.4.5   Experiments and results

For determination of an appropriate number of neurons in the hidden layer of the feed-forward networks we performed small batch set trainings of networks using different numbers of neurons. The result is presented in Figure 1.7.

The empirical test shows that networks are best performing with 40-60 neurons in the hidden layer. We finally decided to use neural networks having one hidden layer with 50 neurons and tan-sigmoid activation functions. These are the neural networks settings used:

- Feed forward neural network (FF) with 20 inputs (normalized values of parameter vectors *P*) and one output (normalized values of C153 marker)
- Pattern recognition neural network (PR) with 20 inputs (normalized values of parameter vectors *P*) and four outputs (four dimensional binary vectors coding classes of C153 marker)
- Feed forward neural network (FC) with 20 inputs (normalized values of parameter vectors *P*) and two outputs (normalized values of C153 marker and normalized classes of C153 with: *Class 1* = 0.25; *Class 2* = 0.5; *Class 3* = 0.75; *Class 4* = 1.0)

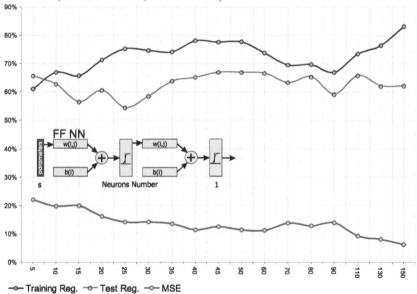

Figure 1.7 Test regression of neural networks with different number of neurons.

All four neural networks were trained with Levenberg-Marquardt algorithm and a validation failure factor 6. The test output of all networks is post processed by aggregation rules based on the decision system and finally compared with original values and classes of C153 tumor marker from test set. The regression functions between the outputs of three networks (PR, FC, FF), post processed final estimation and test C153 values are presented in Figure 1.8. It can be observed that regression of the rules based final estimation of C153 value is greater (R = 0.732) than the individual estimation of separate networks (R= 0.62, R = 0.70, and R = 0.68 for PR, FC, and FF network respectively). The bounded area in the right corner indicates fatal mismatching i.e. originally large values of C153 marker and small-predicted values.

Based on the predicted values of tumor marker C153 we calculate the matching and mismatching ratios for each class of marker. Matching and mismatching of classes between the test data of marker C153 classes and the predicted classes of different networks are presented as confusion matrixes (see

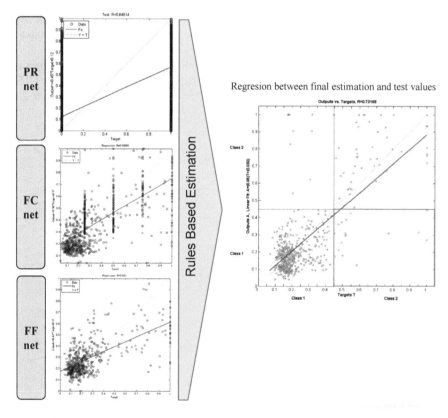

Figure 1.8 Results of Rule Based Estimation System for tumor marker C153.

Figure 1.9). After post processing, we obtain 72 % matching of four classes, whereas the original networks range between 59% and 66% matching cases. The ratio of fatal mismatching is 2.9 %.

Additionally we conducted an experiment with a reduced number of training classes of marker C153. We merge *Class 1* and *Class 2* of marker C153 into a new *Class I* and *Class 3* and *Class 4* into a new *Class II*. That means that all values of tumor marker C153 less than Extreme Normal Value determine *Class I* and values greater than Extreme Normal Value determine *Class II*. Normalization of blood parameters remains unchanged. Full training and test of networks was performed on input data modified in this way. Figure 1.10 presents the confusion matrices for this experiment.

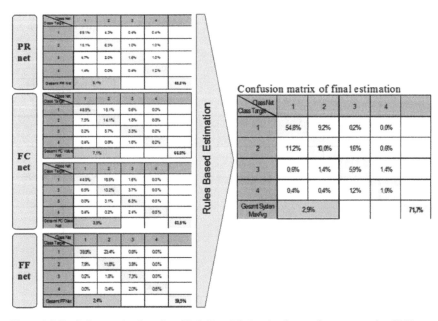

Figure 1.9 Confusion matrix of results of Rule Based Estimation System for tumor marker C153.

Confusion matrix PR NN

Class Net / Class Target	Class I	Class II	
Class I	82,1%	8,9%	
Class II	2,4%	6,5%	
Gesamt PR Net			88,6%

Confusion matrix final estimation

Class Net / Class Target	Class I	Class II	
Class I	84,3%	3,3%	
Class II	2,0%	10,4%	
Gesamt System MaxAvg			94,7%

Figure 1.10 Confusion matrix of two class prediction for tumor marker C153.

We obtain a quite good ratio (94.7 %) of matching cases and the ratio of fatal mismatching is reduced to 2 %.

The dependency of quality of estimation on different composition rules is presented in Table 1.4. It is shown that the *MaxAvg* rule produces the best results.

Additionally, we performed an experiment using vectors containing numbers of measured values P_i from 27 to 15 (maximal 12 missing values allowed), as the number of vectors containing all 27 parameters is very small (about 160 samples). During normalization we replace the missing values with the value -1 thus resulting in input vectors containing all 27 blood parameters. We have obtained 6191 samples as Learning Pattern Set for the neural networks system and 618 independent samples as Test Set.

All three neural networks were trained by Levenberg-Marquardt algorithm using a validation failure value of 6. The test outputs of all networks are post processed by the decision system based on aggregation rules and finally compared with original values and classes of C153 tumor marker from test set.

Table 1.4
Comparison of rules based estimation quality

Rule	Confusion PR Net	Confusion FF Net	Confusion FC Value Net	Confusion FC Class Net	Confusion New Value	Confusion 2 Classes (Class based)	Confusion 2 Classes (Value based)
Max Count	68,2	59,5	66,1	63,5	67,6	93,7	94,5
Max Avg	68,2	59,5	66,1	63,5	**71,7**	**94,7**	**94,7**
MaxMin	68,2	59,5	66,1	63,5	68,2	89,8	94,3
Max Prod	68,2	59,5	66,1	63,5	68,0	90,2	94,5

Table 1.5
Confusion matrixes of final class estimation using input vectors
with missing values and four classes

Class Net Class Target	1	2	3	4	
1	47,1%	13,3%	0,2%	0,2%	
2	8,9%	15,0%	2,3%	0,0%	
3	0,3%	4,5%	3,7%	1,3%	
4	0,3%	0,6%	1,1%	1,1%	
Total System MaxAvg	5,8%				67,0%

Table 1.6
Confusion matrixes of final class estimation using input vectors with missing values
and two classes

Class Net — Class Target	Class I	Class II	
Class I	82 %	4,9 %	
Class II	4,7%	8,4 %	
Total MaxAvg			90,5

Confusion matrixes of final class estimation (Table 1.5) show that the final estimation method works well too with input vectors containing missing values. After post processing, we achieve 67 % matching of four classes, whereas the original networks gets between 53% and 61% matching cases. The ratio of fatal mismatching increases to 5,8 %.

The test results for two classes are shown in a separate confusion matrix (see Table 1.6). We obtain in this case a ratio (90,5 %) of matching and the ratio of fatal mismatching is reduced to 4,7 %.

The regression function between post processed final estimation and test C153 values is presented in Figure 1.11.

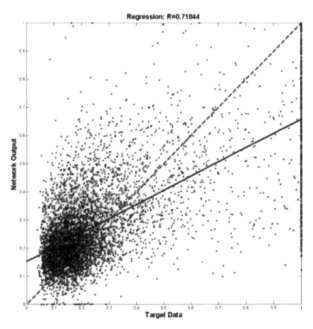

Figure 1.11 Regression functions between post processed final estimation and test C153 values.

It can be expected that not all markers can be predicted with similar performance as C153. The regression between the same blood parameters test data and marker value estimation for tumor marker CEA (regression 0,53) is presented in Figure 1.12.

The results of cancer prediction with additional estimation of missing values of tumor markers in comparison to prediction with incomplete data show that the probability of incorrect diagnosis of positive cancer appearance decreases [6].

Test results (test data include ~ 2400 samples with max. 3 missing values of C-markers) for our case study in breast cancer with the previously shown marker group *C* for both systems is presented in Table 1.7 and Figure 1.13. As can be seen the overalll probability of correct diagnoses decreases but also the percentage of false negatives (negative diagnosis but actually cancerous disease existent).

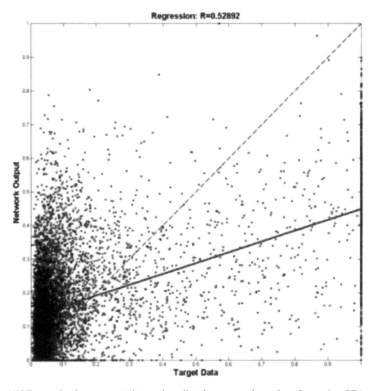

Figure 1.12 Regression between test data and predicted tumor marker values for marker CEA.

Table 1.7
Confusion matrix of breast tumor prediction based on C124, C153, C199 and CEA marker
group without and with blood parameter based marker value estimation.

Neural Networks	P(1/1)	P(1/0)	P(0/0)	P(0/1)	P$_{correct}$
Neural networks system without missing value estimation (vector C as input)	16,4	9,3	58,0	16,3	74,4
Neural networks system with blood parameters based missing value estimation (vector C as input)	21,8	18,9,4	48,4	10,9	70,2

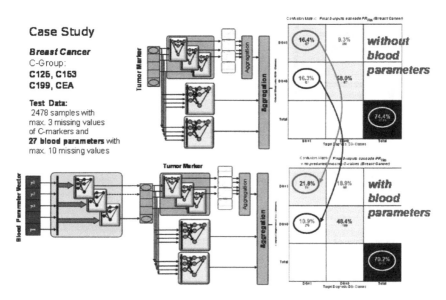

Figure 1.13 Confusion matrix of breast tumor prediction based on C124, C153, C199 and CEA marker group without and with blood parameter based marker value estimation.

1.4.6 System with additional networks for cancer occurrence prediction trained with blood parameter data

As we have observed the introduction of additional information (estimation of missing values) to the system leads to a better prognosis of positive cases of cancer occurrence but decreases the general probability of correct diagnosis. To make use of properties of both approaches we coupled the previously described systems into one system extended by additional networks for cancer diagnosis using only blood parameter data for training. The structure of the whole system is presented in Figure 1.14.

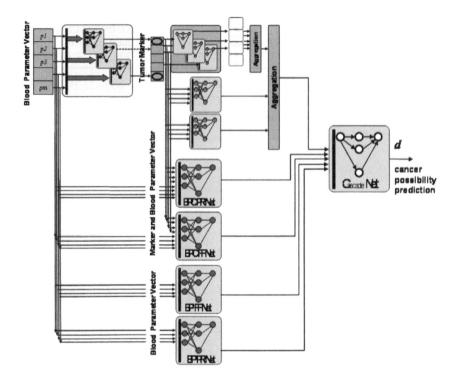

Figure 1.14 Structure of full neural networks based system for cancer diagnosis.

The whole system consists of feed forward and pattern recognition networks using as inputs: the vector of tumor markers **C**, the vector of tumor marker with estimated missing values $C_{estimated}$, the combined vector of tumor marker and blood parameter values **BPC= (C,P)** and the vector **P** of blood parameters only. The outputs of all networks (prediction of cancer occurrence) are used as input for the final feed forward network, which calculates the diagnosis.

Table 1.8
Confusion matrix of breast tumor prediction based on C124, C153, C199 and CEA marker group without and with blood parameter supported neural networks

Neural Networks	P(1/1)	P(1/0)	P(0/0)	P(0/1)	$P_{correct}$
Neural networks system without missing value estimation (vector C as input)	16,4	9,3	58,0	16,3	74,4
Neural networks system with blood parameters and vector C as input	26,4	17,4	49,9	6,3	76,4

1.5 RESULTS

We compare the results of prediction of cancer diagnosis between the system using only tumor marker information and the full system using information coming from tumor marker and blood values. The test data concern breast cancer diagnosis based on four previously described tumor markers and 27 standard blood parameters. The confusion matrices in Table 1.8 and Figure 1.15 show results for both systems using the same test data. It can be seen that the combined

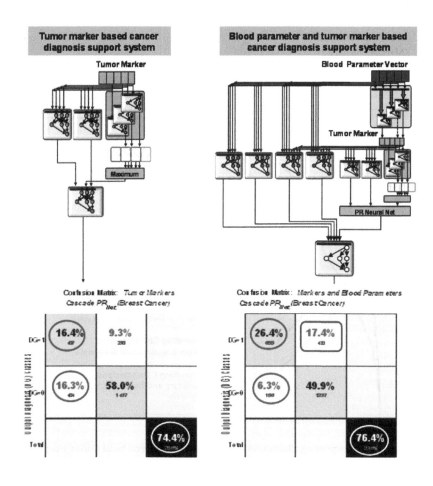

Figure 1.15 Comparison of breast tumor prediction based marker group without and with blood parameter supported neural networks.

blood parameter and tumor marker system:

- increases the probability of correct cancer diagnosis (true positives)
- decreases the probability of incorrect diagnosis of positive cancer appearance (false negatives) but also
- increases the probability of incorrect diagnosis of negative cancer occurrence (false positives)

Generally, the introduction of blood parameters makes the system more pessimistic in respect to cancer prognosis: It predicts a positive diagnosis although a cancer disease does not actually exist

The Receiver Operating Characteristic for both systems obtained for the previously described test is presented in Figure 1.16.

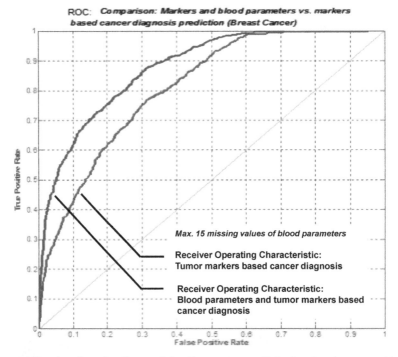

Figure 1.16 Receiver Operating Characteristic of breast tumor prediction based marker group without and with blood parameter supported neural networks.

References

[1] Astion, M.L., Wilding P., 1992, Application of neural networks to the interpretation of laboratory data in cancer diagnosis. Clinical Chemistry, Vol 38, 34-38.

[2] Djavan, B., Remzi, M., Zlotta, A., Seitz, C., Snow, P., Marberger, M., 2002, Novel Artificial Neural Network for Early Detection of Prostate Cancer. Journal of Clinical Oncology, Vol 20, No 4, 921-929

[3] Harrison, R.F., Kennedy, R.L., 2005, Artificial neural network models for prediction of acute coronary syndromes using clinical data from the time of presentation. Ann Emerg Med.; 46(5):431-9.

[4] Jacak W., Proell K., 2010a, Data Driven Tumor Marker Prediction System, *Proceedings of EMSS 2010*, Fes, Marokko

[5] Jacak W., Proell K., 2010b, Neural Network Based Tumor Marker Prediction, *Proceedings of BroadCom 2010*, Malaga, Spain

[6] Jacak W., Proell K, 2011- Neural Networks Based System for Cancer Diagnosis Support - Proceedings of International Conference on Computer Aided Systems Theory EUROCAST 2011, Las Palmas, Spanien, , pp. 283-285

[7] Liparini, A., Carvalho, S., Belchior, J.C., 2005, Analysis of the applicability of artificial neural networks for studying blood plasma: determination of magnesium on concentration as a case study. Clin Chem Lab Med.; 43(9):939-46

[8] Matsui, Y., Utsunomiya, N., Ichioka, K., Ueda, N., Yoshimura, K., Terai, A., Arai, Y., 2004, The use of artificial neural network analysis to improve the predictive accuracy of prostate biopsy in the Japanese population. Jpn J Clin Oncol. 34(10):602-7.

[9] Penny, W., Frost, D., 1996, Neural Networks in Clinical Medicine. Med Decis Making, Vol 16: 386-398

[10] Reibnegger, G., Weiss, G., Werner-Felmayer, G., Judmaier, G., Wachter, H., 1991, Neural networks as a tool for utilizing laboratory information: Comparison with linear discriminant analysis and with classification and regression trees. PNAS: vol. 88 no. 24, 11426-11430

Chapter 2.

Myoelectric Control of Upper-Limb Prostheses and the Effects of Fatigue

Stefan Hermann and Klaus Buchenrieder

Universität der Bundeswehr München, Germany
stefan.herrmann@unibw.de; klaus.buchenrieder@unibw.de

Advanced upper-limb prostheses are commonly controlled with myoelectric signals (MES), which are directly obtained from the skin-surface. The measured MES is filtered and pre-processed before pattern recognition and classification methods are applied, to uncover prehensile motions or gestures. To match patterns within the MES to hand positions, Linear Discriminant Analysis (LDA) techniques are commonly used. One example for such a classification algorithm is the Guilin-Hills Method developed at the Universität der Bundeswehr München (UniBw). It is based on the observation, that conditional probability density functions over features of the myoelectric signal are normally distributed, exhibit identical class covariances and have full rank. Under these conditions, the Bayes-rule for quadratic discriminant analysis allows to classify hand positions based on previously computed classification maps.

2.1   INTRODUCTION

In 1912 Piper was one of the first scientists who discovered the electromyographic signal (EMS) [1]. In his work, he measured cutaneous voltage changes with a string galvanometer, recorded the signals and described the properties. The first myoelectric control mechanism was implemented by Reiter [2] who developed a myoelectric prosthetic arm, designed for an amputee factory worker [3]. The first working prototype was demonstrated by Reiter at the Hannover Export fair in 1948. Even though, the device was rather bulky and not portable, the idea behind the

Bio-Informatic Systems, Processing and Applications,   27-52,

control system was fundamental. The MES from a single contracting muscle of the residual limb was amplified to control a wooden hand, which was actuated by an electric solenoid. The control for opening and closing actions was derived from different rhythms of contraction. Today this is known as a rate-control system, in which short-duration or long burst signals correspond to open- or close-actions. Today state-control systems, in which bursts are used to initiate changes or selections between motion, speed or joint action are often used, in contrast to Reiter's original scheme.

Reiter's system closed the hand when no MES was detected for a predefined period of time. While the hand was open, a MES with a lower amplitude caused the hand to close gradually. This combination of voluntary opening with automatic closing is known today as cookie-crusher control. Unfortunately, Reiter's invention required sophisticated electronics not available then and only in the late 1950's researchers began re-inventing parts of Reiter's myoelectric control system. Simultaneous work, conducted in Europe, the USSR, Canada and in the USA, was mainly due to the invention of the transistor which made portable MES controlled prosthesis possible.

Advances in the UK by Battye and Nightingale in 1955 [4] show, that two muscles, producing the movement in normal operation, are needed to control hand opening and closing. All prostheses today rely on this result which ensures precise control over the actuator. The first clinically successful, battery powered prostheses, was developed in the USSR and the fitting rights sold to international groups. It is important to note, that, the USSR Academy of Sciences and the Central Prosthetics Research Institute in Moscow introduced the concept of dimension reduction through features, calculated from the MES. In their work they also note, that parameters drawn from the frequency characteristics would prove useful in the future [5].

These results were futile until Swedish researchers began looking at signal processing techniques in 1964. In 1965, the division of Applied Electronics at Chalmers University of Technology in Göteborg began research in the area because they believed devoutly in improving prosthetic control methods with algorithms. This fundamental work prepared ground for more accurate controls of myoelectric systems based on pattern recognition, information theory and adaptive components. Weltman and Lyman at UCLA introduced an "Autonomous Control System capable of supplementing the operator's own conscious control by generating future end-point positions using the Maximum Likelihood decision rule to accept or reject a predicted position based on past movements performed by the system" [6].

Advances in microelectronics, signal processing and microcontroller systems have led to numerous EMG controlled applications. Most prominent are flight-control applications [7], electronic music systems [9], control devices for exoskelettal robots [9] and user interfaces for portable electronics in general [10]. A consummate survey of historical developments and the state-of-the-art in the field is provided by Merletti and Parker [11]. An essential compendium of advances in the design and the technology of prostheses was compiled by Näder in 1988 [12]. A more recent survey of scientific and practical progressions, concerning upper-limb prosthesis and microprocessor control was collocated by Lake and Miguelez [13]. Accessorily, Muzumdar [14] provides an extensive overview over excitation methods, prosthetic implementations and clinical applications in his general survey book.

The fundamental classification problem for prostheses-control was outlined and structured by Englehart and his affiliate researchers, who established the foundation for interdisciplinary research in this field [15]. The researchers describe the classification challenge as a multistage process consisting of: a feature extraction-, a dimensionality reduction- and a classification-phase, as depicted in Figure 2.1. The MES is obtained with active bipolar skin-surface electrodes and unwanted noise and dc-voltages are suppressed or cancelled with a differential instrumental amplifier. The amplification is followed by a bandpass filter and an amplitude adjustment. The resulting signal is then presented to the feature extraction stage. Multichannel MES are considered as high dimensional data vectors within predefined fixed time-intervals. Describing the signal using a smaller set of data reduces calculation time and classifier complexity. For this purpose, features are calculated as originally suggested by Kobrinski and his colleagues. Feature vectors are used as input for training or classification purpose. Depending on the classification scheme, features may be extracted in the time-, in the frequency-(Fourier transform) or in the time-frequency-domain (Short-Time Fourier Transform, Wavelet- or Wavelet Packet-Transform).In this chapter, we introduce the Guilin-Hills method and restrict the presentation for clarity to the standard time-domain elements: Mean Absolute Value (MAV), Root Mean Square (RMS), Waveform Length (WFL), Willison Amplitude (WAMP), Slope-Sign Changes (SSC), Integrated Absolute Value (IAV), Variance of the myoelectric signal (VAR) and Zero Crossings (ZC).

In the dimensional reduction phase, the space of the initial set of features is reduced to retain important information from the signal and to discard unimportant data. This helps to achieve the commonly required 300 msec timespan for classification so that the patient will not notice a time delay. There are two strategies: feature projection and feature selection. Feature projection methods generally yield the best combination of original features through a mapping of the original multidimensional feature-space into a space with fewer dimensions. Typically an original feature-space is transformed with a linear transformation, e.g., with methods, such as principal component analysis (PCA). PCA provides a means of unsupervised dimensionality reduction, as no class membership qualifies the data when specifying the eigenvectors of maximal variance [15]. In contrast, feature selection methods provide the best subset of the initial feature set. In the context of myoelectric signal processing, feature selection often relies on Euclidian distance calculations and class separation criteria. Supervised methods rank the features according to the available class membership information, so that the "centers of gravity" for the sample-clouds of hand-positions, obtained during, i.e., a training phase, are maximally distant.

Based on Engleharts' and Hudgins' research on the established multistage classification scheme, myoelectric control methods are widely accepted. Today, most modern upper-limb prostheses rely on Hudgins' theoretical foundations using a multistage classification. A significant number of methods employ a multilayer perceptron as classifier with time-domain features as input data [15, 16, 17]. Another large group of conventional prosthetic devices, like the Otto Bock Sensorhand, is controlled by the amplitude or the rate of change of the MES 15]. Even though amplitude dependent schemes appear simple at first, these prove very robust and accurate and are successfully used by many amputees [15]. With only two sensors, one axis can be controlled at a time as originally demonstrated by Battey and Nightingale. With an additional enhancement, the co-contraction, users can switch between control-axis or between different modes of operation. Pattern classification schemes, i.e., the Guilin-Hills Selection Classifier offer more flexibility, in that it is possible to control more than one axis at a time. With our Guilin-Hills classification algorithm we are able to instantly distinguish up to nine different hand positions without a need for mode changes.

Unfortunately all classification methods or algorithms are prone to fail or degrade on the onset of muscle fatigue [18]. This fact is often neglected when control algorithms are developed for prostheses or other MES controlled systems.

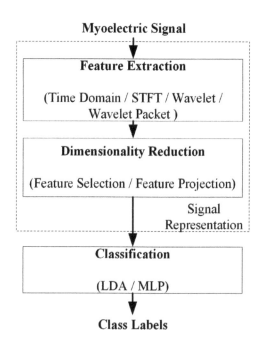

Myoelectric Signal

Feature Extraction

(Time Domain / STFT / Wavelet / Wavelet Packet)

Dimensionality Reduction

(Feature Selection / Feature Projection)

Signal Representation

Classification

(LDA / MLP)

Class Labels

Figure 2.1 Schematic visualization of the multistage myoelectric classification problem [15].

Muscle fatigue is commonly detected and monitored by a down-shift of features expressing the mean-frequency [19, 20, 21], the median-frequency [22], or the ratio of low-frequency to high-frequency components [23]. In comparison to frequency-domain effects, the behavior of time-domain features and their distribution, especially the consequence on pattern classification algorithms, is not fully understood. The main reason for the change of the myoelectic signal under the influence of fatigue, is the variation of the muscle conduction velocity [24, 25]. Physiologists attribute this effect to changes in electrochemical metabolite concentrations within the tissue surrounding the muscle [26]. In addition, the firing rate of the participating muscle fibers decreases as the force-twitches of motor-units ascents [27]. As a result the dispersion of feature values increases.

Since the classifier introduced in this chapter solely relies on time-domain attributes, we mainly focus here on time-related features and calculate the fatigue-dependent classification rate. To maintain a high classification accuracy for sustained muscle contractions, the classifier must clearly be able to compensate for fatigue [27]. Since features from all domains of the myoelectric signal are affected only near optimal solutions can be achieved.

2.2 FEATURES OF THE MYOELECTRIC SIGNAL

In muscle fiber packs, motor units are coinstantaneously triggered over an axon, but the neuro-muscular junctions of the fibers are physically distributed and thus, fascicles of a muscle are triggered asynchronously. For this reason, a constant muscle contraction results in time-distributed overlapping bursts of Motor Unit Action Potentials (MUAP). The nature of the observable motor unit discharge appears highly irregular [28]. Merlo et al. [29] model a MES as:

$$
\begin{aligned}
s(t) &= \sum_j MUAP \cdot T_j(t) + n(t) \\
&= \sum_j \sum_t k_j \cdot f\left(\frac{t - \theta_{t,j}}{\alpha_j}\right) + n(t)
\end{aligned}
\tag{2.1}
$$

where k_j is the amplitude factor for the jth Motor Unit and $f\left(\frac{t - \theta_{t,j}}{\alpha_j}\right)$ denotes the shape of the action potential discharge. The term $\theta_{t,j}$ designates the occurrence time of the MUAP, α_j represents a scaling factor, and $n(t)$ describes additive noise. Generally, a MES is considered as a complex, non-stationary stochastic signal formed by the superimposition of numerous, individual action potentials, generated by irregular discharges of active Motor Units in the muscle. Consequently, an MES must be mapped into a smaller dimension vector, called either a feature vector or just feature. In calculating features, a recorded MES is segmented into equidistant, consecutive frames of, i.e., 256 msec each.

Researchers frequently segment the signal into overlapping windows to smooth the results and to generate more classifier outputs per second.

For the example classification algorithm, the Guilin-Hills method, time-domain features from the standard time-domain elements are selected, as succinctly described in [30, 31].

2.2.1   Root Mean Square (RMS)

The Root Mean Square, also known as the quadratic mean, is a statistical measure for the magnitude or strength of the muscle contraction. The RMS for N samples $\{x_i, \cdots, \square_n\}$ is:

$$RMS = \sqrt{\frac{1}{N} \sum_{i=1}^{N} x_i^2} \qquad (2.2)$$

2.2.2   Waveform Length (WFL)

The Waveform Length feature is useful to classify steady muscle exertions and indicates the frequency range plus the amplitude of the signal for a muscle contraction:

$$WFL = \sum_{i=1}^{N} |x_i - x_{(i-1)}| \qquad (2.3)$$

2.2.3   Zero Crossings (ZC)

Zero-Crossings denote the count of sign-changes of a signal within a predefined time-window. The threshold S ensures noise reduction. The zero crossings criterion summarizes frequency related properties. For two consecutive samples x_k and x_{k+1}, over the entire time frame holds:

$$ZC = \sum_{k=1}^{N-1} g_{zc}(x_k) \quad \text{with} \qquad (2.4)$$

$$g_{zc}(x) = \begin{cases} 1 & \text{if } sgn(-x_k \cdot x_{k+1}) \wedge (|x_k - x_{k+1}|) \geq S \\ 0 & \text{otherwise} \end{cases} \qquad (2.5)$$

whereby

$$sgn(x) = \begin{cases} true & \text{if } x \geq 0 \\ false & \text{otherwise} \end{cases} \qquad (2.6)$$

2.2.4   Mean Absolute Value (MAV)

The Mean Absolute Value provides a quantity for the intensity and the dynamic change of a muscle contraction:

$$MAV = \bar{X} = \frac{1}{N}\sum_{i=1}^{N}|x_i| \tag{2.7}$$

2.2.5   Integral Absolute Value (IAV)

The Integral Absolute Value is generally viewed as a measure to distinguish strong and weak contractions of the probed muscle. It can be used to appoint the speed for the motion of the prostheses. It calculates:

$$IMAV = MAV \cdot N \tag{2.8}$$

2.2.6   Willison Amplitude (WAMP)

The Willison Amplitude is again a criterion for the excerted force of the muscular contraction. It counts how often the difference of consecutive sample values exceeds a threshold:

$$WAMP = \sum_{i=1}^{N-1} f(|x_i - x_{i+1}|) \tag{2.9}$$

$$f(x) = \begin{cases} 1 & if\ x > threshold \\ 0 & else \end{cases} \tag{2.10}$$

2.2.7   Variance (VAR)

The Variance scales the statistical dispersion of the MES and averages the squared distance of its values from the expected value:

$$VAR = \sigma^2 = \frac{1}{N-1}\sum_{i=1}^{N}(x_i)^2 \tag{2.11}$$

2.2.8   Slope Sign Changes (SSC)

The Slope-Sign Changes-measure provides information about the base frequency of the MES. As for the ZC feature, the threshold S reduces noise effects:

$$SSC = SSC + 1;\quad if\ (x_k - x_{k-1}) \cdot (x_+ - x_{k+1}) \geq S \tag{2.12}$$

Characteristics of sustained contractions are rated with spectral features from the frequency- and from the time-frequency-domain. Muscle fatigue influences the amplitude as well as the shape and the scaling factor over time. Features in all domains are affected. Changes in the frequency-domain however, are most prevailing and thus, researchers often neglect the effects in the time- and in the time-frequency domain [32]. For quick assessment and monitoring of muscle-fatigue, researchers often monitor the mean frequency (MNF), the median frequency (MDF) and the mean scale (MNS) [31, 33].

2.2.9   Mean Frequency

The Mean-Frequency feature is defined as the first-order moment for the average frequency of the power spectrum [22].

$$MNF = \frac{\int_0^\infty \omega \cdot P(\omega) \cdot d\omega}{\int_0^\infty P(\omega) \cdot d\omega} \tag{2.13}$$

Where $P(\omega)$ is the power spectrum density (PSD) of the myoelectric signal and the frequency is represented by ω.

2.2.10   Median-Frequency

The frequency, at which the spectrum is divided into two parts of equal power, is considered the Median-Frequency (MDF). With the zero-order moments of the PSD follows:

$$\int_0^{MDF} P(\omega) \cdot d\omega = \int_{MDF}^\infty P(\omega) \cdot d\omega = \frac{1}{2} \int_0^\infty P(\omega) \cdot d\omega \tag{2.14}$$

2.2.11   Mean Scale

Oskoei, Hu and Gang [22] use the features Mean-Scale (MNS) and the inverse MNS, also known as the Instantaneous Mean-Frequency (IMNF), to analyze fatigue in localized dynamic contractions. In their contributions, they rely on the continuous wavelet transform, which is defined as:

$$CWT(s, \tau) = \int x(t) \Psi_{s,\tau}^*(t) \cdot dt \tag{2.15}$$

With the scale parameter s, translation or time-shift parameter τ, and the MES $x(t)$. The basic function $\Psi_{s,\tau}^*(t)$ is a scaled and shifted version of the mother wavelet $\Psi(t)$:

$$\Psi_{s,\tau}^*(t) = \frac{1}{\sqrt{s}} \Psi\left(\frac{t-\tau}{s}\right) \tag{2.16}$$

The power density function or scalogram (SCAL) is estimated [34] with:

$$SCAL(x, \tau) = |CWT_x(x, \tau)|^2 \qquad (2.17)$$

Like the Mean-Frequency, the Mean-Scale is defined as the first-order moment of the scalogram:

$$MNS = \frac{\int_{ls}^{hs} s \cdot SCAL(s) \cdot ds}{\int_{ls}^{hs} SCAL(s) \cdot ds} \qquad (2.18)$$

Thereby, ls is the lowest and hs the highest scale of interest.

2.3 THE UNIBW-HAND

Based on the foundations established by Englehardt and Hudgins, several prostheses control systems were successfully developed at the Universität der Bundeswehr München. The first UniBw-Hand prototype relied on a classical multilayer perceptron with distinct modes for training and operation. In more recent work, a data-glove was introduced to improve the classification results. The custom data-acquisition glove is fitted over the able-hand to complement the MES sensors, which are placed over the residual muscles on the opposite arm. During the training phase probands are asked to perform identical hand-motions on both sides so that the obtained weights for the neural net precisely relate to the intended gestures. Even though significant improvements of the neural-net based classifier were achieved through guided teaching, further improvements concerning repeatability, robustness and accuracy would not be achieved. This led to the development of a different classifier scheme.

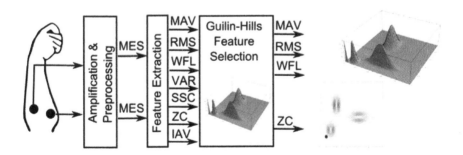

Figure2.2 To ensure highly reliable classification results optimal feature muscle combinations are selected. They are chosen so that the classification regions have the least overlap.

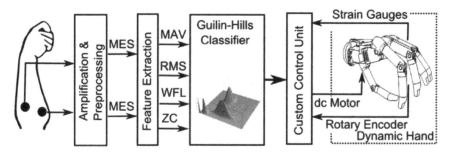

Figure 2.3 The UniBw-Hand control system for powered upper-limb prostheses control. The Guilin-Hills Classifier inputs features and issues hand position information to the motor control unit.

Starting from the Bayes theorem, the Guilin-Hills Classification method for the UniBw-Hand was developed [35]. Just as in the neural-net approach, the operation of the new system is also divided into a training-based feature selection- (Figure 2.2) and an operative-phase (Figure 2.3). A proper subsets of the original feature space is derived and selected during the training-phase, so that a minimal number of criteria relate a hand-position to feature-muscle pairings.

The pairings are calculated from the MES, gained from the residual predominant muscles, e.g., on the patients' forearm, for predefined hand and arm motions or positions. In case the correlation of a hand-position and a muscle-feature pairing is ambiguous, additional pairings are calculated to resolve the equivocality. In the last step of the training-phase, the two-dimensional probability density functions are transformed into classification-maps for a quick table-lookup like process in the operative-mode.

While the prostheses is in operation, the preprocessing process of the MES is exactly as in the training-mode. The previously selected normalized muscle-feature combinations are passed to the classification-map to obtain the correct hand position. The lookup is quasi real-time and the motor driving the gripper of the prostheses reacts without a noticeable time-delay. The custom motor control unit of the UniBw-Hand is the interface to the dynamic hand and contains a current limiter for the motor and dedicated sensors which provide feedback to the control unit ensuring a force regulated and slippage reduced grip.

2.4 THE GUILIN-HILLS METHOD

The Guilin-Hills Selection and Classification Method is a statistical cluster analysis technique and based on the Bayesian classification theory. It combines the advantages of established dimensionality reduction methods, such as feature-selection and feature-projection. The common goal of dimensionality reduction techniques is to retain important information for best class discrimination and to discard superfluous information [36, 37]. Feature selection methods select a subset of features according to a set of criteria from an original feature set. In contrast, feature projection methods strive to determine the best combination of the initial features to achieve a new feature set through mapping or combination of features into a lower dimensional space [31].

The Guilin-Hills Selection Method [35], [30] is based on the observation, that time-domain features and some of the features in the frequency-domain are normally distributed within the feature space. Even though this observation is not self-evident at first, it was confirmed by testing several hundred training data sets, collected from different individuals. All data-sets were found to be mesokurtic, with $Y_N \approx 0$. Consider the four distinct hand-positions: radial flexion, wrist-extension, fist and radial-flexion depicted with normal distributions in Figure 2.4.

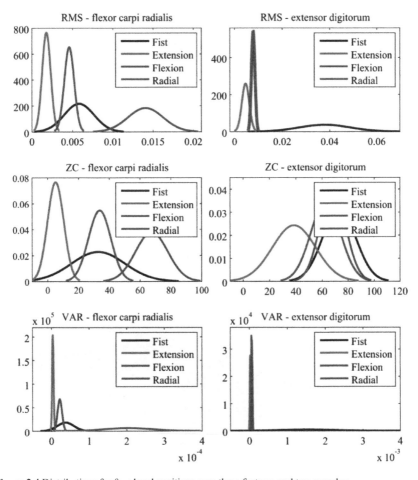

Figure 2.4 Distributions for four hand-positions over three features and two muscles.

In the graphs, the features: Waveform Length (WFL), Root Mean Square (RMS) and Zero-Crossings (ZC) are shown for the muscles: Flexor Carpi Radialis and Extensor Digitorum. Each of the distributions represents one hundred training samples for a hand-position. Since the distributions overlap, a single feature is clearly not sufficient to discern the individual hand-positions. For this reason, at least two features must be combined to unambiguously discriminate between hand-positions or gestures. The combination of two muscle-feature pairings, sufficient to distinguish four hand-positions, is shown in Figure 2.5. Even though it was found, that two dimensions are sufficient to distinguish seven hand-positions, additional dimensions can be added to enhance the classifier.

The Guilin-Hills Selection Method involves two phases: a training-phase, in which feature-muscle combinations are selected for the construction of classification-maps, and an operative-phase, in which the previously calculated maps are used to classify MES for prostheses control. Distinct phases are advantageous, because only the first phase involves computationally expensive operations, whereas the calculations in the second phase can be carried out in real-time with an energy efficient micro- or signal-processor embedded in the prostheses.

During the training-phase, all features are calculated from the measured MES and combinations of features are selected. Each combination is expressed with a

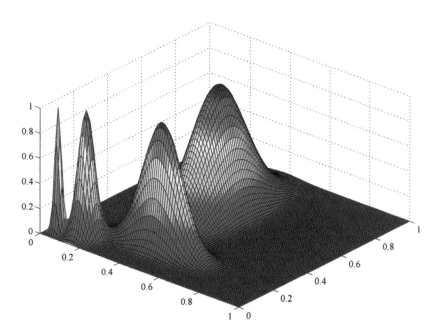

Figure 2.5 Normalized two-dimensional probability density functions for the hand-positions: fist, wrist-extension, wrist-flexion and radial-flexion over the features X=RMS1 and Y=RMS2.

normalized probability density function, representing a hill:

$$PDF_{A,B}(x,y) = \frac{1}{2\pi\sigma_A\sigma_B\sqrt{1-\rho^2}} \cdot e^{\left(-\frac{1}{2(1-\rho^2)}\left[\left(\frac{x-\mu_A}{\sigma_A}\right)^2 - 2\rho\frac{x-\mu_A}{\sigma_A}\cdot\frac{y-\mu_B}{\sigma_B} + \left(\frac{y-\mu_B}{\sigma_B}\right)^2\right]\right)} \tag{2.19}$$

Where ρ refers to the correlation coefficient between the random variables X and Y with the expected values μ_A and μ_B and the standard deviations σ_A and σ_B. It follows from:

$$\rho = \frac{Cov(A,B)}{\sigma_A\sigma_B} = \frac{E\big((A-\mu_A)(B-\mu_B)\big)}{\sigma_A\sigma_B} \tag{2.20}$$

where E is the expected value operator and Cov means covariance. Correlation coefficients are directly derived from the training data and determine the angular orientation of the hill.

An unambiguous association of a MES with a position or a hand-motion is only possible, when all probability density functions are clearly separated without an overlap. The separation is calculated and expressed by the worst-case crossover (WCC) measure. It represents the maximal joint-equal probability for adjoining normalized probability density functions. Two hills that inherit a small WCC are well separated.

$$WCC(\mu_1,\sigma_1,\mu_2,\sigma_2,\rho_{12},\mu_3,\sigma_3,\mu_4,\sigma_4,\rho_{34}) =$$

$$Min\left[\frac{e^{\left[\frac{\left(\frac{1}{\sigma_1^2}+\frac{(\mu_2-\mu_4)^2}{\sigma_2^2(\mu_1-\mu_3)^2}+\frac{2\rho_{12}(\mu_4-\mu_2)}{\sigma_1\sigma_2(\mu_1-\mu_3)}\right)(\mu_1-x)^2}{2(1-\rho_{12}^2)}\right]}}{2\pi\sigma_1\sigma_2\sqrt{1-\rho_{12}^2}} + \frac{e^{\left[\frac{\left(\frac{1}{\sigma_3^2}+\frac{(\mu_2-\mu_4)^2}{\sigma_4^2(\mu_1-\mu_3)^2}+\frac{2\rho_{34}(\mu_4-\mu_2)}{\sigma_3\sigma_4(\mu_1-\mu_3)}\right)(\mu_3-x)^2}{2(1-\rho_{34}^2)}\right]}}{2\pi\sigma_1\sigma_2\sqrt{1-\rho_{12}^2}}\right]$$

$$\tag{2.21}$$

for:

$$peak(\mu_1,\sigma_1,\mu_2,\sigma_2,\rho_{12}) < x < peak(\mu_3,\sigma_3,\mu_4,\sigma_4,\rho_{34}) \tag{2.22}$$

By calculating the WCC for all possible muscle-feature combinations, the hills which are maximally distant, are found. Even though, a few hundred combinations may exist, the number of combinations is still small enough to obtain optimal classification maps through an exhaustive search in an acceptable amount of time. In addition, the complexity of the search can be reduced when unfeasible combinations are eliminated, e.g., when a crossover exceeds a threshold value and thereby reveals immoderately amalgamated hills. Ideally, the result of the search provides a single muscle-feature combination in which all hills are perfectly separated. Because such a single combination does generally not exist for more than two positions, we compute a tree of contour-plots instead. Figure 2.6 shows four hand positions for which two overlapping regions exist. Within these regions an unambiguous classification is not possible. With a spy-glass-like process, the ambiguities are resolved in that alternative contour-plots provide disjoined footprints. As an example, consider the normalized feature maps for the RMS of sensor1 and 2, nrms1-nrms2, at the top in Figure 2.6. Here the footprints for the hand-positions: radial flexion and wrist flexion overlap. In case, a measurement falls within the covering, the detailing nrms1-nzc1 feature map is chosen instead. The measurement will then palpably point to the correct hand-position – the ambiguity issue is resolved. Discrete classification maps, containing only identifiers for hand-positions, gestures or pointers to detailing digital maps, are generated in the last step of the training phase.

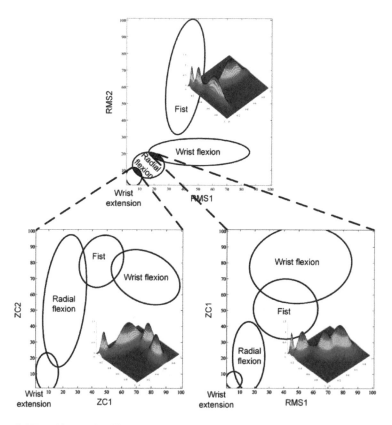

Figure 2.6 Unambiguous classifier outputs can be resolved by linking to different classification maps, where the classes in question are better separated.

The introduction of discrete maps is advantageous, because subsequent classification steps can be carried out with a lean computing device. Note, that the maps can be held in flash memory and updated whenever required by the user or by an orthopedics technician.

During the operative phase of the prostheses, features are continuously extracted from the MES and normalized for categorization with the classification-maps. Thereby, the classification is reduced to an iterative process that can be carried out in real-time.

2.5 EFFECTS OF FATIGUE

The fundamental assumption of the Guilin-Hills Selection Method is, that a constant muscle contraction leads to time-domain features, for which mean-value and standard-deviation do not significantly vary over time. For long lasting muscle contractions this assumption is false, after the onset of fatigue. Five right handed participants, with an average age of 25 years and no known muscular disorders, were invited for an experiment to evaluate fatigue effects. The participants were asked to perform a fist at maximum voluntary contraction for a duration of 30 seconds followed by a relaxation phase. For each volunteer, data for a dozen hand poses were acquired, processed and evaluated.

In the setting, we used a Delsys Bagnoli Desktop EMG System with four channels and DE3.1 Double Differential Sensors. The Delsys signal-acquisition system offers a variable amplification and has a built in bandpass filter with a lower cutoff frequency of 20 Hz and a upper cutoff frequency of 450 Hz.

For the experiments, an electrode was placed over the Extensor Digitorum muscle on the right forearm of the volunteers. The EMS was digitized with a National Instruments PCI 6251 DAQ Card. The data was sampled at 1024 Hz incorporating a custom build software, guiding the participants through the acquisition. Raw data was stored and features were calculated afterwards with MatlabTM Version 7.7. In the sessions, several different window sizes and window overlap values were determined for traces of fatigue. With a fixed window size of 256 samples and no window overlap, several threshold values for the WAMP, ZC and SSC features were evaluated. To compare the results of our experiments with findings of other researchers [21], the mean frequency of the acquired data was compared.

2.5.1   Myoelectric Features and Fatigue

The experiments clearly show, that fatigue significantly affects all features in the time-, the frequency- and in the time-frequency domain. Since the probability density functions of all features change when fatigue starts, not only our Guilin-Hills selection method, but all Linear Discriminant Analysis classification algorithms without dynamic compensation are strongly affected. It was also found, that the degree of the effects varies significantly among participant.

Effects on Time Domain Features

The experiments show a variation of the standard-deviation and a decrease in the mean-value of all time-domain features after the onset of fatigue within the 30 second test intervals. The inanition starts after 5 to 10 seconds and the determined scaling factor spans from 0.16 to 0.95. As an example, consider Figure 2.7. It shows the probability density function of the WFL features. Each of the PDF is calculated for a three second time-frame. After 20 seconds, there is no overlap with the distribution representing the first three seconds of the training data. Consequently, a classifier incorporating a distribution, based on short training data sets, provides fewer correct classifications during longer periods of muscle exertion. Summarized experimental results are provided in Table 2.1 through Table 2.3.

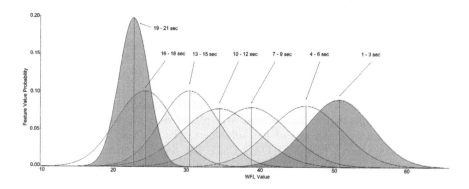

Figure 2.7 Changes of the probability distribution for WFL over time, due to muscle fatigue.

Table 2.1

Comparison of RMS feature decrease between the start and the end of the experiment for different participants

Participant	Start value	End value	Percentage
1	0.2179	0.115	52.7
2	0.2294	0.1141	49.7
3	0.2308	0.2244	97.2
4	0.2341	0.1473	62.9
5	0.0926	0.0736	79.5

Table 2.2

RMS feature values of one participant with different window sizes and no overlap between consecutive windows

Window-Size	Start value	End value	Percentage
64	0.2144	0.1134	52.9
96	0.2156	0.1142	52.9
128	0.2165	0.114	52.7
192	0.2176	0.1137	52.3
256	0.2179	0.115	52.8
320	0.2181	0.1143	52.4
384	0.2183	0.1246	57.1
448	0.2184	0.1248	57.1
512	0.2184	0.1276	58.4

Table 2.3

RMS Feature values of one participant with different window overlaps and a window size of 256 samples

Window Overlap	Start value	End value	Percentage
0	0.2179	0.115	52.8
32	0.2178	0.1138	52.2
64	0.2177	0.1168	53.7
96	0.2166	0.1174	54.2
128	0.2174	0.1151	52.9
160	0.2169	0.1157	53.3
192	0.217	0.1162	53.5
224	0.2168	0.117	54

Although the data generally follow the expected pattern of decreasing MES amplitude, the variance for the five probands is significant.

Table 2.1 2.1 shows the change of the RMS mean values, spanning from 50% to 3% for different individuals. Hence, simple means of classifier adjustment are not viable. Also note, that the effect of changing the window-size adjustment for the calculation of the features or the respecting overlap is of little significance. Table 2.2 presents the negligible change of the RMS features over window-sizes from 64 to 512 samples. Likewise, window-overlap does not provide insight into the progress of fatigue. Table 2.3 shows the RMS feature values for one participant at a fixed window size of 256 samples.

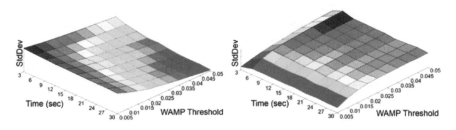

Figure 2.8 Mean and Standard Deviation for WAMP and respective changes over time..

The threshold values of the WAMP, SSC and ZC features have a major impact on the significance of the observed feature value. Figure 2.8 visualizes the mean-value for WAMP and the standard-deviation over time for different threshold values. The observed fatigue and the decreasing factor clearly depends on the chosen threshold. As the experiments show, threshold selection is crucial, because low values lead to saturation; since excessive noise is detected even under the presence of fatigue, showing almost no decline of the mean feature value. Higher threshold values produce correct values with the drawback of a fatigue dependent decay. Consequently, thresholds must be carefully chosen since classification accuracy and threshold values are interdependent.

Effects on Frequency- and Time-Frequency-Domain Features

The mean-frequency is one of the standard features to detect and monitor fatigue. For all our participants a decrease in the mean-frequency was observed. Figure 2.9 shows one example of the mean-frequency, of the power spectrum within

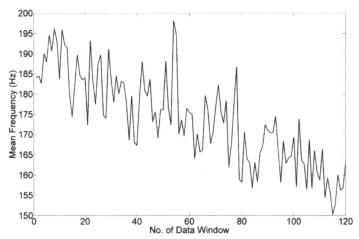

Figure 2.9 Mean frequency for 30 second of maximum voluntary muscle contraction, visualizing a decreases of about 12 percent.

windows, 256 samples wide, for a 30 second experiment. The mean-frequency at the end of the trial is about 12 percent down from the initial value. The mean-scale is also an approved feature to observe and rank muscle fatigue. As an example refer to Figure 2.10, which provides a representative mean-scale and a wavelet scalogram.

The visualizations show, that the amplitude of the signals reduce over time. This effect is also given in the scalogram and the mean scale, however it can not be clearly seen, because too many samples and noise effects cover the trend. For this reason, the average mean scale, shown in the graph below, depicts the reduction.

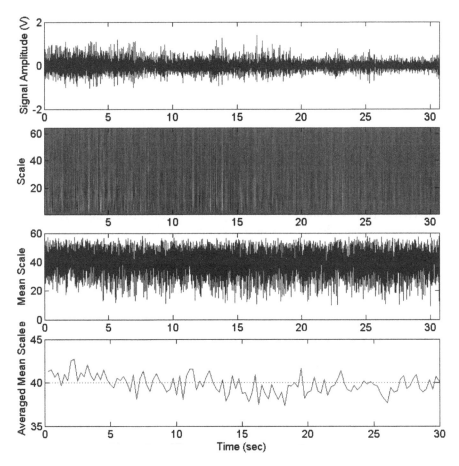

Figure 2.10 Visualization of the changes in the myoelectric signal, the wavelet scalogram, the feature mean scale and the averaged mean scale.

2.5.2   Missclassification Probability

A measure to determine how fatigue effects classification methods is needed. Therefore the probability of misclassification over time was estimated. For each experiment the initial three seconds were used to produce a classification map. This three seconds are equal to the time that is normally used during the training phase. Because of fatigue, which is progressing with time, the features shift and void correct classification. The initial classification map is described through the following probability density function, with A and B representing applied features with the mean values μ_A, μ_B, the standard deviation values σ_A, σ_B, and the correlation coefficient ρ.

$$PDF_{A,B}(x,y) = \frac{1}{2\pi\sigma_A\sigma_B\sqrt{1-\rho^2}} \cdot e^{\left(-\frac{1}{2(1-\rho^2)}\left[\left(\frac{x-\mu_A}{\delta_A}\right)^2 - 2\tilde{n}\frac{x-\mu_A}{\sigma_A}\cdot\frac{y-\mu_B}{\sigma_B}+\left(\frac{y-\mu_B}{\sigma_B}\right)^2\right]\right)}$$

(2.23)

For the Guilin-Hills method, the classification area is defined through the 5% footprint of the probability density distribution and is expressed by:

$$Class_{A,B}(x,y) = \begin{cases} 1, if \ e^{\left(-\frac{1}{2(1-\rho^2)}\left[\left(\frac{x-\mu_A}{\sigma_A}\right)^2 - 2\rho\frac{x-\mu_A}{\sigma_A}\cdot\frac{y-\mu_B}{\sigma_B}+\left(\frac{y-\mu_B}{\sigma_B}\right)^2\right]\right)} \geq 0.05 \\ 0, otherwise \end{cases}$$

(2.24)

Integrating the probability density function equals to the probability of the observed variables. It follows, that the probability of successfully classifying a myoelectric signal after a given time t_x can be calculated via the probability for which a feature value-tuple remains within the original classification area. It is calculated through the integral of the pdf at the observation time multiplied with the original classification function:

$$P_{suc,t_x} = \int_{x=-\infty}^{\infty}\int_{y=-\infty}^{\infty} PDF_{t_x,A,B}(x,y)\cdot Class_{A,B}(x,y)\cdot dx\cdot dy$$

(2.25)

This misclassification probability can be considered as a direct measure for the reliability of the classification algorithm. Figure 2.11 shows the probability, for which the combination of the features RMS and WFL remain within the footprint of the multinormal distribution at the beginning of the experiment. After 9 to 12 seconds, only three of four feature tuples are correctly classified. After 30 seconds, two third of all classification results represent the wrong hand position. This is clearly insufficient for prosthetic control. The experiments show, that the probability of misclassification significantly varies for different feature combinations and different participants. Misclassification is affected by the feature mean value decrease, standard deviation value and its variation, which is patient dependent.

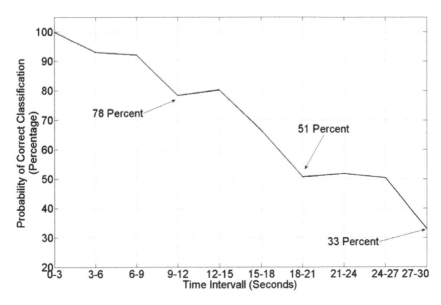

Figure 2.11 Decreasing probability of a positive classification for a fist, based on RMS and WFL features.

2.6 CONCLUSIONS

Classification algorithms based on Linear Discriminant Analysis often rely on the assumption, that myoelectric features are normally distributed. This assumption however, is only valid until the onset of muscle fatigue. Thereafter calculated features are not normally distributed anymore resulting in the failure of stationary classification schemes. Figure 2.12 visualizes the histogram of the VAR feature values for several 30 second experiments and a normal distribution plus a generalized extreme value distribution for the data. Due to the skewness of the data set, the generalized extreme value distribution is preferable Using the two dimensional feature maps, generated by the Guilin-Hills Selection Method for fatigued signals, hand-positions are not correctly categorized. This can be improved with dynamic compensations, e.g., maps which correct the shifts during prolonged muscle exertions, as an example refer to Figure 2.13. The footprint of a two-dimensional normal distribution appears shifted with respect to the original training data set.

Major findings of the work in progress concerns the general loss in accuracy, which effects all pattern recognition classifiers relying on Linear Discriminant Analysis (LDA) with time-domain, frequency-domain and time- frequency-domain features, due to fatigue. With statistical methods, it is possible to determine and single out sensitive features to achieve better classification results.

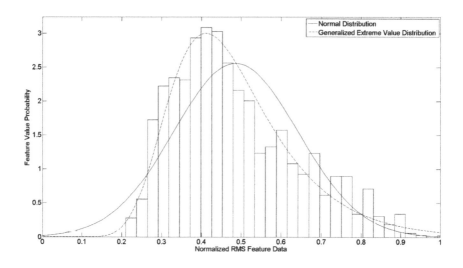

Figure 2.12 Normalized RMS feature value histogram and the corresponding normal and generalized extreme value distribution.

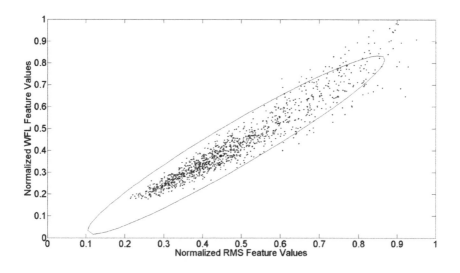

Figure 2.13 Distribution of RMS and WFL feature values for several 30 second trials and the corresponding 5 percent footprint of the classification map.

Also, the significance of features varies over time and good results in the beginning may become inadequate over time. This effect is currently investigated considering long time reliability. Since it is necessary to distinguish between voluntary feature decrease and fatigue, this task is not trivial and the mere tracking of features is not sufficient.

Current work targets a system at which characteristics of the MES are acquired during the training phase, so that fatigue effects can be compensated when the prostheses is in use.

Such indicators will enable us to dynamically adjust the classification regions, according to the amount of fatigue. Because all features calculated from the myoelectric signal are affected by fatigue, we currently develop a new hybrid sensor, which combines myoelectric and near infrared spectroscopy sensors in one device. This hybrid enables researchers to assess fatigue by the detection of changes in the hemoglobin concentration in close proximity of the acting muscle. During a measurement, light is induced to the tissue and the amount of backscattered photons indicates muscle contraction. The sensor integration offers the opportunity for better classification results.

2.7 QUESTIONS

1 On what behalf is a dimensionality reduction introduced in the classification scheme for prostheses control?
2 What is the advantage of splitting the myoelectric signal classification using the Guilin-Hills classifier into a training- and a operation-phase?
3 What is the reason for representing classification areas as digital maps?
4 Why are all Linear Discriminant Analysis based classification schemes affected by muscle fatigue?
5 What is the reason for calculating features for myoelectric signals?
6 Name the advantage of pattern recognition techniques over common control techniques like the amplitude- or rate-coding.

References

[1] H. Piper, Electrophysiologie menschlicher Muskeln. Berlin: Springer-Verlag, 1912.
[2] Pudulski, "The Boston Arm," Forum IEEE Spectrum 6, 1969.
[3] R. Reiter, "Eine neue Elektrokunsthand," Grenzgebiete der Medizin, vol. 4, pp. 133-135, 1948.
[4] C. Battye, A. Nightingale, and J. Whillis, "THE USE OF MYOELECTRIC CURRENTS IN THE OPERATION OF PROSTHESES," J. Bone Joint Surg., vol. 37B, pp. 506-510, 1955.
[5] A. Kobrinski, S. Bolkovitin, and L. Voskoboinikova, "Problems of

bioelectric control," in Proceedings of the 1st IFAC International Congress of Automatic and Remote Control, Vol. 2, Butterworths, London, 1960, pp. 619-620.

[6]   A. Freedy and J. Lyman, "Adaptive Aiding," in Proceedings of the 3rd International Symposium on External Control of Human Movement, Dubrovnik, 1969, pp. 155-170.

[7]   C. Jorgensen, K. Wheeler, and S. Stepniewski, "Bioelectric Control of a 757 Class High Fidelity Aircraft Simulation," Technical Report: Computational Sciences Division, NASA Ames Research Center, vol. http://ase.arc.nasa.gov/publications/pdf/1999-0108.pdf (seen Jan/20/2008), 1999.

[8]   F. Aubin and J.-M. Robert, "Bioelectric input devices: an example: BIOLINK," in ACM Conference on Human Factors in Computing Systems, Proc. 1992 SIGCHI Conference on Human Factors in Computing, Monterey, CA, 1992, pp. 95-96.

[9]   C. Fleischer, A..Wege, K. Kondak, and G. Hommel, "Application of EMG signals for controlling exoskeleton robots," Biomed Tech, vol. 51, pp. 314-319, 2006.

[10] W. Hill, F Pereira, Y Singer, and L. Terveen, "Wireless myoelectric control apparatus and methods," 6244873, 1999.

[11] R. Merletti and P. Parker, Electromyography - Physiology, Engineering and Non Invasive Applications. Hoboken, NJ: John Wiley & Sons, 2004.

[12] M. Näder, Prothesen der oberen Extremität.: Mecke Druck und Verlag Duderstadt, 1988.

[13] C. Lake and J. Miguelez, "Comparative Analysis of Microprocessors in Upper Limb Prosthetics," Journal of Prosthetics and Orthotics, vol. 15, pp. 48-65, 2003.

[14] A. Muzumdar, Powered Upper Limb Prostheses, Control, Implementation and Clinical Application.: Springer Verlag Berlin Heidelberg, 2004.

[15] K. Englehart, B. Hudgins, P.A. Parker, and M. Stevenson, "Classification of the Myoelectric Signal using Time-Frequency Based Representations," Special Issue of Medical Engineering and Physics on Intelligent Data Analysis in Electromyography and Electroneurography, vol. 21, pp. 431-438, 1999.

[16] B. Hudgins, A New Approach to Multifunction Myoelectric Control, 1991, Ph.D. Thesis, (New Brunswick, Canada, Department of Electrical Engineering, University of Frederiction).

[17] B. Hudgins, P. Parker, and R. Scott, "A New Strategy for Multifunction Myoelectric Control," IEEE Transactions on Biomedical Engineering, vol. 40, pp. 82-94, 1993.

[18] D. MacIsaac, P. Parker, K. Englehart, and D. Rogers, "Fatigue Estimation with a Multivariable Myoelectric Mapping Function," IEEE Transactions on Biomedical Engineering, vol. 53, pp. 694-700, 2006.

[19] F. Molinari, M. Knaflitz, P. Bonato, and M.V. Actis, "Electrical

Manifestations of Muscle Fatigue during Concentric and Eccentric Isokinetic Knee Flexion-Extension Movements," IEEE Transactions on Biomedical Engineering, vol. 53, pp. 1309-1316, 2006.

[20] L. Gilmore and C. DeLuca, "Muscle Fatigue Monitor (MFM): Second Generation," IEEE Transactions on Biomedical Engineering, vol. BME-32, pp. 75-78, 1985.

[21] S. Karlsson, J. Yu, and M. Akay, "Enhancement of Spectral Analysis of Myoelectric Signals During Static Contractions using Wavelet Methods," IEEE Transactions on Biomedical Engineering, vol. 46, pp. 670-684, 1999.

[22] M. Oskoei, H. Hu, and J. Gan, "Manifestation of Fatigue in Myoelectric Signals of Dynamic Contractions Produced During Playing PC Games," in Proc. 30th Annual International Conference of the IEEE Engineering in Medicine and Biology Society EMBS 2008, 2008, pp. 315-318.

[23] F. Stulen and C. DeLuca, "Frequency Parameters of the Myoelectric Signal as a Measure of Muscle Conduction Velocity," IEEE Transactions on Biomedical Engineering, vol. BME-28, pp. 515-523, 1981.

[24] S. Gandevia, "Spinal and Supraspinal Factors in Human Muscle Fatigue," Physiol Rev, vol. 81, pp. 1725-1789, 2001.

[25] R. Merletti, S. Roy, E. Kupa, S. Roatta, and A. Granata, "Modeling of Surface Myoelectric Signals. II. Model-Based Signal Interpretation," IEEE Transactions on Biomedical Engineering, vol. 46, pp. 821-829, 1999.

[26] D. MacIsaac, P. Parker, R. Scott, K. Englehart, and C. Duffley, "Influences of Dynamic Factors on Myoelectric Parameters," IEEE Engineering in Medicine and Biology Magazine, vol. 20, pp. 82-89, 2001.

[27] C. DeLuca, "The Use of Surface Electromyography in Biomechanics," Journal of Aplied Biomechanics, vol. 13, pp. 135-163, 1997.

[28] M. Oskoei and H. Hu, "Myoelectric Control Systems - A Survey," Biomedical Signal Processing and Control, vol. 2, pp. 275-294, 2007.

[29] A. Merlo, D. Farina, and R. Merletti, "A Fast and Reliable Technique for Muscle Activity Detection from Surface EMG Signals," IEEE Transactions on Biomedical Engineering, vol. 50, pp. 316-323, 2003.

[30] K. Buchenrieder, "Processing of Myoelectric Signals by Feature Selection and Dimensionality Reduction for the Control of Powered Upper-Limb Prostheses," Computer Aided Systems Theory - EUROCAST, vol. LNCS 4739, pp. 1057-1065, 2007.

[31] M. Zecca, S. Micera, M. C. Carrozza, and P. Dario, "Control of Multifunctional Prosthetic Hands by Processing the Electromyographic Signal," Critical Reviews in Biomedical Engineering, vol. 30, pp. 459-485, 2002.

[32] P. Bonato, S. Roy, M. Knaflitz, and C. DeLuca, "Time-Frequency Parameters of the Surface Myoelectric Signal for Assessing Muscle Fatigue During Cyclic Dynamic Contractions," IEEE Transactions on Biomedical Engineering, vol. 48, pp. 745-753, 2001.

[33] K. Englehart, "Signal Representation for Classification of the Transient

Myoelectric Signal," University of New Brunswick, Ph.D. theses 1998.

[34] N. Radicheva, L. Gerilovsky, and A. Gydikow, "Changes in the Muscle Fibre Extracellular Action Potentials in Long-Lasting (Fatiguing) Activity," European Journal of Applied Physiology and Occupational Physiology, vol. 55, pp. 545-552, 1986.

[35] K. Buchenrieder, "Dimensionality Reduction and Classification of Myoelectric Signals for the Control of Upper-Limb Prostheses," Proceedings of the IASTED - Human Computer Interaction 2008, pp. 113-119, 2008.

[36] S. Theodoridis and K. Koutroumbas, Pattern Recognition, Fourth Edition.: Academic Press, 2008.

[37] I. Fodor, "A survey of dimension reduction techniques," Report of the U.S. Department of Energy. Report Nr.: UCRL-ID-148494, vol. Lawrence Livermore National Laboratory, 2002.

Chapter 3.

Using Game Consoles for Human Medical Data Collection: in-field applications

Christopher Armstrong[1], Diarmuid Kavanagh[1], Peter Rossiter[2], A/Prof Sara Lal[1]

[1]*University of Technology, Neuroscience Research Unit, School of Medical and Molecular Biosciences, Sydney, Australia*
[2] *Forge Holdings Pty Ltd, Sydney, Australia*
SARA.LAL@UTS.EDU.AU, PMR@FORGE.COM.AU

3.1   INTRODUCTION

In this chapter, we consider and discuss how it may be possible to cost effectively and reliably support biomedical research in-the-field. Biomedical research is now challenged with the need to conduct more of its research outside the traditional laboratory, this in-the-field biomedical research may need to be conducted in the participant's home, work place or any environment in which humans engage. This type of research is only now becoming a practical possibility for most researchers due to the increasing availability of Information and Communications Technology (ICT) and access to semi-autonomous computer systems that can collect such data and send it back to the laboratory for processing and dissemination.

We will consider the selection of the required software and hardware components as key aspects of this type of research. Systems supporting biomedical research in-the-field need to be reliable, allow for considerable customizability and convenient accessibility while also remaining secure and easily locked down against unwanted access.

A set of requirements for hardware and software for such systems will be explored. We then utilize these requirements to evaluate the use of game console as a hardware platform with the appropriate software and make useful comparison to other hardware choices.

Bio-Informatic Systems, Processing and Applications, 53-68,

To complete the outline, we consider network and data transmission, data security and reliability before finishing with a summary.

Through a number of ongoing projects and one in particular associated with the smart application of physiological and psychological data to the trucking industry (SmartData project), the broad aim is to study the effects of environmental and human factors and how they relate to risk factors in the work place. A recurring need for reliable and semi-autonomous gathering of biomedical data in-the-field was identified. The purpose of the SmartData project, which we will use as an example, is to build hardware and software infrastructure that will allow profiles of drivers to be compiled through continuous non-intrusive collection of physiological and psychological data in real time. The aim is to close the gap that remains in the study of physiological and psychological factors and fatigue. The effects of fatigue on drivers has been examined extensively through laboratory-based research ([1]; [2]), but important road environment field research still remains to be done and has so far proven difficult to undertake due to lack of appropriate data collection technology.

While many of the effects of fatigue can be studied in the laboratory, there are limits on the ability to collect realistic ecological data and the acceptability of laboratory results for drawing conclusions about real on-road driver risk factors needs to be verified by more field research. This is not the only instance of the need to verify laboratory-based studies by equal field-based research. Much of the human factors research is waiting for a convenient, practical and cost effective solution to instrumentation and data acquisition for in-the-field use. A possible solution starts with a practical technology for collecting data about participants outside the laboratory, preferably from their own homes, places of work or even vehicles in the case of truck drivers. If we take truck drivers for example, such technology would need to enable the capture of factors that may influence on-road driving risk that may be present prior to driving long distances, factors at home or in truck depots. Robust and reliable technology for continuous real time monitoring of physiological and psychological factors in the field would provide more ecologically valid data sets reflective of the true physiology and psychology of the sample group. In the case of truck drivers, the ability to acquire and measure important data prior to long driving episodes can take current research to new areas of exploration.

It turns out that the requirement for a low cost reliable in-the-field biomedical data acquisition and management technology is a recurring theme. There may be a possible extension of our research into the investigation of home-based care applications for the aged through constant round the clock non-intrusive physiological and psychological monitoring ([3]; [4]; [5]). Hence, there is a need to find a cost effective hardware and software combination that could stand up to the rigor of field deployment and at the same time provide a reliable and robust platform on which to deploy and support ongoing scientific investigations.

There are many examples in the literature over the last 10 years of investigations in the field of telemedicine and telecare ([3]; [4]) and all require an underlying hardware and software data acquisition combination for biomedical data acquisition. Most usually involve the use of a PC or laptop as these are common and easy to obtain, however, as we will illustrate they do come with limitations and still leave many questions of robust deployment and management unanswered. Indeed, this paper suggests that one of the issues that has hampered the emergence from the laboratory to the home of many of the ideas and innovations investigated in the field of telemedicine and telecare is the lack of a cheap and reliable instrument for performing numerous and varied field investigations particularly in homes.

The home is a very hostile and unpredictable environment in comparison to the laboratory and it is difficult to manage successful deployment of standard PC-based solutions because they are easily accessible and vulnerable to modification and re-configuration. The standard PC and laptop with well known operating systems invite interference by both expert and novice.

We are unaware of any suitable devices available in the market that could be customized to such research needs. Therefore, this paper describes the development of a 'gateway' device, which may be used as a hub to download data from medical monitoring devices, gather information from psychological surveys and schedule physiological tests, all in the home environment. The ability to remotely monitor and control the 'gateway' device is also an important consideration.

This paper considers game consoles (such as the Sony PlayStation 2 and 3, Microsoft Xbox and Microsoft XBox 360) as hardware platforms for the gateway device. We use the term 'gateway device' to indicate the use of a dedicated computer that is placed in a person's home, and is used to coordinate a number of other digital and electronic devices that are connected to it since this is the start of a flexible biomedical data acquisition and management technology.

The programming language Java with an OSGi framework was chosen as our software platform because of its suitability for residential gateway devices and relative portability and security ([6]; [7]; [8]). This paper will further proceed to describe experiments undertaken in order to satisfy the base requirements with two game consoles, (i) the Sony PlayStation-2 (Sony Computer Entertainment, PLAYSTATION 2, Australia) and (ii) Sony PlayStation-3 (Sony Computer Entertainment, PLAYSTATION 3, Australia) and the relative strengths and weaknesses of each console.

Much of our previous work and our future investigations are focusing on the popular notion of the smart home and [3] [4] and there are many reasons for developing a cheap, reliable and resilient hardware and software combinations that can be deployed to the field and in particular the home environment.

3.2 REQUIREMENTS

The SmartData project aims to collect data with relatively simple, off-the-shelf electronic devices that monitor heart rate, physical activity, etc., and examine this in the context of self-reported results from participant's response to psychological surveys (such anxiety, emotion, mood etc.) and certain physiological tests (reaction time, ECG etc.). However, the data from such devices must be constantly downloaded onto a desktop computer, and the effort required to manually collate and analyze this data using conventional methods such as paper-based forms makes such research substantially time-consuming and expensive. Furthermore, once a research program has been started, it can be very difficult to change aspects of it during its progression, especially if the participants are geographically displaced from the main research site for long periods of time, as would be in the case of truck drivers.

It is apparent that a device is required that could be placed in a participant's home to act as a hub or gateway to collect data and feed it back to the researchers in a central location. This "gateway device" would allow the participants to provide response to electronic surveys. In order to monitor participants' progress without physical access to the gateway device for long periods (sometimes up to a few weeks), the device would need to be controlled remotely, able to download updates, as well as subsequently upload research data after the research program commences.

It was also a requirement that the device be readily available and easily obtainable. Even better if the device was a commodity while not an immediate or interesting target for hackers or denial of service attacks. It has been noted in the literature that as devices for collecting important physiological and psychological data become more prevalent and critical to quality of life then the consequences of connectivity to the Internet and the inherent dangers of hacking and denial of service increase [5].

It was also a requirement that the device should be able to work with a minimal of support equipment. That is, it should be self contained and able to work to a standard TV on the basis that all homes are likely to have a TV set. The device should be ambitious in the sense that it can be readily deployed worldwide to all field situations, so no legitimate scientific investigation is hampered by lack of electronic resources and does not place a burden on potential participants and scientific endeavor to source additional monitors or anything but the standard power supply available in the local area.

The use of a PC as a solution was ruled out because it is too hard to control the operating system versioning, software updates and numerous other malicious pieces of software which can find their way onto the PC when connected to the Internet. As it is a requirement to connect to the Internet, standard operating systems like Microsoft XP or Vista are too hard to maintain in a known configuration. Often Microsoft itself will install updates and changes in the background unless this feature is disabled by the user, however, it is to easily re-enabled. It is also too easy for participants and others to interfere with these common operating systems. A piece of instrumentation for scientific investigation needs to be as secure as possible and in a known state of configuration to ensure reliable measurement.

It was logical to look for a more resilient operating system and a less well known platform that was less inviting and less well known and therefore through obscurity protected from many of the configuration management issues encountered with PC-based hardware.

3.2.1 Hardware

The above requirements give us a basic architecture of the proposed system, which would be similar to that in Figure 3.1.

The type of software required to collect the data from such external devices and to allow updates to program schedules, is specialized, and was developed by the SmartData project for this purpose. We wanted to reduce the time needed for software testing during development by requiring that the target device support a full Java Standard Edition runtime environment and not have the software satisfy stringent performance requirements. This means that the device cannot be a resource-constrained system and should be capable of running a 32-bit, pre-emptive, multitasking operating system with virtual memory management and memory protection. The device had to be cheap to be deployed in large numbers but customizable enough to permit the SmartData software to run. There was a strong desire for the device to be able to connect to participant's television sets to reduce the costs associated with purchasing separate display units. A simple way to connect the monitoring devices using the Universal Serial Bus (USB) is a required because many of off-the-shelf sensors and monitors that need to be worn or carried by the participant whilst they are not home, such as an activity monitoring device like the Actigraph GT1M Monitor, which collects and records body movements [9] come with USB connectivity. Large data sets would be derived so the device needs secondary storage. It is also important to be able to lock down the device so that participants could not misuse the device or interfere with its normal operation.

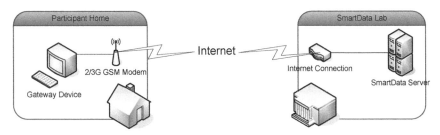

Figure 3.1 Architecture of SmartData system.

3.2.2   Software platform

As this box will be deployed "out in-the-field" and has to be kept running for long periods, a software platform with high reliability that could be updated whilst the system is still running is required. Software with relative portability was also desirable in order to deploy the same or similar software on later gateway devices to be modified for deployment in trucks as well as to collect 'real field' driving data. We wanted flexibility to change hardware platforms and operating systems where we deemed it to be necessary. Java was chosen over languages such as C/C++ because it tends to be more portable, provides fine granularity of in-process security through the Java Permissions Application Programming Interface (API), and it's lack of direct pointer-access can help reduce common programming errors that are fatal (such as null pointer exceptions and dangling pointer dereferences) and cause security issues such as buffer overflow exploits [10].

Java served as a base for the OSGi platform, which was designed to run on residential gateway devices that have continuous uptime and need to be managed and updated remotely, without user intervention. It extends Java to provide a structure that can be used to build inter-changeable software components that are loosely coupled in the form of services, allow fine-grained security configurations, and can be updated without restarting the framework [11]. There are multiple vendors that implement the specification ([12]; [13], [14]; [15]; etc). This gives more flexibility in choosing an environment that suited our research needs and makes it easier to switch implementations later if required.

There was also a preference to get Java running on the platform in order to use the PlayStation-2 as a deployment and server environment for wireless sensor motes (namely the TMote Sky and TMote Invent, Sentilla, USA) as Java support was easier to work with for development of these motes. It was intended to use these motes to perform tests, such as a reaction time assessment test.

3.2.3   Selecting appropriate hardware

We examined multiple computer-type devices suitable for our needs. We first looked at an emerging field of devices used such as home gateways which are Personal Video Recorders (PVRs) for watching and recording broadcasted television (TV) program and Set Top Units for watching digital TV and subscription TV services. They are mass-produced and built to interact with a TV, which makes them suitable for our application. Unfortunately, many are tied to proprietary platforms without much scope for modification, such as set top units distributed with cable-TV services, and others are not on the market for long enough so that we can make a firm decision.

PC-based systems were also considered as they are considerably cheap and have good software and peripheral support due to widespread use. Apart from being able to support a greater range of operating systems (such as Windows,

Linux, etc.) there is also good support for development due to their widespread use. It has excellent backwards compatibility, and there are a wide variety of vendors that produce PC-based hardware, making it very difficult to be "locked-in" to one manufacturer. It is also possible to source hardware boards that have most peripherals integrated.

The main concern with PCs was how easily they could be used by the participants to install another operating system, whereas dedicated devices tend to require expert technical skills in order to modify them. There is also considerable variance amongst all the PC-based products in the market with most products having short lifecycles before a new variant is introduced, making it harder to predict performance if the hardware needs to be frequently changed. Using newer hardware means that we would need to thoroughly test new software drivers, which in our experience are more unstable in newer versions. This would be a more practical option in the future if we could select products with long lifecycles, and a simple method could be identified to "lock-down" the devices.

The last option being considered is game consoles. Similar to PCs, they are relatively cheap due to mass production and remain compatible with their software over the product lifecycle. Their product lifecycles are quite long, for example, the PlayStation One was available in Japan from December 1994 [16] and discontinued from March 2006 [17]. PlayStation 2 was introduced in March 2000 [16] and continues to be available, and the Microsoft Xbox was introduced in March 2000 [18] and discontinued in 2006. They also connect easily to TV sets. Customization is a big concern as most consoles are proprietary and like set top units are not easily modifiable. There is a risk of vendor "lock-in" because only one company or organization typically manufactures each model. The only commercially available consoles identified to allow customization without purchasing prohibitively expensive Software Development Kits were Sony PlayStation-2 and PlayStation-3, both of which can run Linux with fairly inexpensive modifications supported by the manufacturer. The Xbox (Microsoft Xbox) was also a consideration as it has been shown to run Linux, but it required a legally dubious and unsupported "mod-chip" in order to boot Linux and changes to its controller ports in order to access its USB hub. We ruled this option out because we wanted minimal hardware changes and less uncertain legal issues.

3.2.4   PlayStation 2

The PlayStation-2 (PS2) features multi-core MIPS architecture called the "Emotion Engine" (clocked at about 290MHz) with cores designed for handling image processing, video-decompression and vector calculation [19]. There are separate processors for handling graphics, audio and PlayStation-1 games. It has about 32MB of RAM, composite TV-OUT or RF-OUT, two controller ports with memory card slots, and two USB 1.1 ports at the front.

Because of these powerful hardware features, many technically advanced purchasers of the PS2 expressed an interest in running Linux on their consoles.

Due to the copyright protection implemented by Sony to prevent pirated games from being executed in the console, it was not possible to simply create a bootable Linux disc, plug in a hard drive, and install Linux. The cost of licensing an official development kit for the PS2 is also prohibitive enough so that only commercial game and application developers can afford to purchase one.

Sony published a "PlayStation-2 Linux Kit" (Sony Computer Entertainment, Sony Playstation 2 Linux Kit) that was cheap enough for purchase by consumers. The kit included a hard disk, mouse and keyboard, a disc for booting GNU/Linux, another disc containing a Red-Hat based GNU/Linux distribution, a computer monitor adaptor and an expansion card containing a network card and hard disk connector for attaching the hard disk. The kit requires older hardware versions of the PS2 that contains an expansion slot for plugging in the add-on card (specifically models before the SCPH-7000), which is something that does not appear on the newer 'slim' hardware and makes it impossible to use the PS2 Linux kit [20] with newer hardware.

This kit contains versions of Linux and GNU software that is outdated by today's standards (most sourced from the time the kit was developed), running the Linux kernel 2.2 and a GNU Compiler Collection (GCC) 2.95. This made it more difficult to cross-compile and run newer software and more difficult to use some of the newer hardware devices (such as the Actigraph) that we intended to use. As the PlayStation-2 is a specialized hardware platform with a unique variant of MIPS architecture and peripherals, the official Linux and GNU maintainers do not provide support.

Most of the software required for the research was compiled directly on the PS2 hardware or by using a cross-compiler that was obtained from the PlayStation-2 Linux Community website [20].

The PS2's CPU architecture lacks two useful instructions, 'll' and 'sc', which are used together to implement atomic swap-and-load instructions for simple synchronization primitives. These instructions are used extensively in some open source Java virtual machines for synchronization between threads instead of kernel-based locks. For synchronization that only needs to last for a few CPU cycles (such as incrementing an integer value), this type of locking is more efficient as it avoids the overhead of a system call and context switch into supervisor mode. It was determined not worthwhile to modify the virtual machines and adapt them for system-call based locking because these user-space locks were found to be used extensively in their source code. Furthermore, it would seriously degrade performance by adding many more system calls for simple synchronization needs. The kaffe open-source Java virtual machine [21] was the only one found to have been specifically made compatible for the PS2, whilst others such as jamvm and cacaovm still relied on these specific locking instructions found in normal MIPS architectures.

Another major problem identified was with software performance. Most of the Java virtual machines trialed did not support Just-In-Time compilation for the MIPS architecture, or if they did (in the case of kaffe), they did not have the support for the PlayStation-2's unique architecture. This meant we had to turn on the slower, C-based interpreter for all the virtual machines trialed. This meant that the performance was unacceptably slow and unable to even start up the OSGi framework due to the small amount of RAM and slow execution speed.

Assessing other virtual machines such as jamvm yielded mixed results. The compiler tool-chain being used was too old (GCC 3.x) to support newer code. We did consider the option of modifying the source code of these programs so that they could be compiled with the older GCC, however, this was beyond the team's level of expertise.

One of our more significant reasons for not continuing with the PS2 was the lack of available hardware. The Linux Kit is only available for some regions now (having sold out in the United States and other places) and no longer includes the add-on card needed for network access and to plug in the hard disk; it only has the DVDs and VGA cable. Because it requires the older, larger PS2 and the discontinued expansion port, both these parts need to be carefully located and purchased second hand. This means that there is no guarantee that those parts will be readily available to carry out the proposed research. Even if the slim PS2 models with the cut-down Linux Kit could be used, there is no room or place to connect and mount a hard disk inside the unit.

The main support for the kit was obtained from the PlayStation-2 Linux Community Web Site [20]. Amateur software developers managed to port the PlayStation-2's specific changes to a GCC 3.3 and a Linux 2.4 kernel and published their results on this website, but due to a lack of developer resources, they have not been able to port the patch sets to more recent GCC and kernel versions (Linux 2.6 and GCC 4.x). Additional updates to the kit from Sony for newer software versions on the PS2 could not be found.

The official development kit was also not a viable option, not only because of the cost, but also because an agreement would need to be executed with Sony who would have to approve the software we wanted to produce and would force us to print our own discs. The costs involved with the option were considered too impractical for our project and so we decided not to pursue it.

3.2.5   PlayStation 3

The PlayStation 3 (PS3) has a multi-core PowerPC-based Cell architecture designed for intensive multiprogramming and graphics/vector calculation. It was built for complex three dimensional computer games that need to perform intensive physics and graphics calculations. It also has 256MB of general purpose RAM and a further 256MB of graphics RAM. Unlike the PS2, the PlayStation-3 comes with a built-in hard disk drive, Bluetooth, Bluetooth technology, wireless controllers, USB 2.0 ports and 802.11b/g Wi-Fi Wireless Networking adaptor [22].

The graphics output on the PS3 is also more flexible. The PlayStation-2 required a PAL (Phase Alternating Line) or NTSC (National Television System

Committee) compatible TV set or it was difficult to find computer monitors that supported required "sync-on-green" [20], but the PS3 has HDMI (High Definition Multimedia Interface) and composite outputs. This allowed the connection of a digital computer monitor via an HDMI-to-DVI adaptor as well as a High-Definition or normal analogue television (PAL and NTSC).

GNU/Linux and other operating systems are supported natively on the PlayStation-3 via the "OtherOS" facility in the System Dashboard. This is set up by loading a bootloader image file from a CD, DVD or USB flash drive with the PS3 Dashboard, which copies it into internal flash memory and uses it to boot-strap the other operating system.

Similar to the PS2 RTE, Sony has implemented protection of the graphics hardware through a "hypervisor", which sits between the PlayStation-3 hardware and the OtherOS. It still allows access to the processor cores and main RAM, but only provides a framebuffer for graphics output, not full graphics acceleration as it does for games [22].

Running Java virtual machines on the PlayStation-3 has been much more successful. IBM produces a pre-compiled version of a Java 1.5 and 1.6 virtual machine for its PowerPC architecture that runs directly on the PS3 ([23]), and jamvm is also available. We had no problems using them to start up an OSGi environment with the Knopflerfish and Equinox OSGi implementations.

We were able to download and install most of the software we required using pre-compiled packages for the distribution that we were using (most packages came from the PowerPC variant of each distribution). The PS3's cell chip had enough power to compile and run the Java VM and full-profile OSGi framework too. Each distribution run a newer GCC 4.x variant so there was no significant compilation issues.

The above makes the PlayStation-3 a more attractive option, and it appears that Sony supports continued work on the Linux kernel and some user-space utilities for running Linux on the PS3 [22]. At this point in time, it is possible to compile unmodified versions of the latest Linux 2.6 kernel for the PlayStation-3. However, it isn't clear how long Sony will continue to support Linux on the PS3, and they could easily withdraw their support in the future. If Sony decided to do this, we would have to reconsider the other hardware platforms for our gateway device, such as PCs.

Initially, the PS3 was not a compelling option as it is more expensive than the PS2. Even after pricing the extra hardware and software that needed to be purchased for the PlayStation-2 to run Linux, the PS3 is still more expensive compared to the PS2. The PS3 is not expected to become much cheaper for some years. However, given the relative difficulty of finding the necessary PS2 hardware, compared with immediate retail availability of a PS3 system, the extra cost may be justified.

3.3 NETWORK CONSIDERATIONS AND DATA TRANSMISSION

Another requirement of the technology was the ability to handle large scale deployments cost effectively. To conserve both the bandwidth and network time consumed by the gateway device and the processing required by the server, the gateways only connect to the server as clients on an as-needed basis, usually periodically. Once a connection is established, data is transferred in both directions and then the gateway disconnects itself from the Internet. This ensures that the resources are only used as needed.

Connecting in this manner however presents its own issues. This method is not designed for anything other than non-real time information analysis. While the gateways are receiving real time data the central server will not receive this data until the gateways establishes a connection, which depending on the implementation can be several days. This becomes an issue if the server requires current data from all of the gateways or if a critical sensor image update or software update is ready for the gateways.

Messages that are required to be sent across the network are buffered on both the gateways and the server until a communication link is established. Once a link exists the messages are transferred using a transaction mechanism to ensure data delivery is guaranteed through unreliable links. If confirmation is received, data is removed from the messaging queue, else the message is retransmitted when possible.

Currently, there is no direct mechanism to deal with the situation where the server wishes to communicate with the gateway. With 3G wireless technology, the gateways are allocated dynamic Internet Protocol (IP) addresses and the server cannot contact the gateway if the IP address is not know or if the gateway has not connected to the Internet. An out-of-band external notification system is required to inform the gateway of a server connection request which forces the gateway to establish a connection.

The implemented solution was a Short Message Service (SMS) notification system whereby the server could send out SMS as required to fixed numbers attached to the gateway devices. If possible the gateway would establish the connection to the server over the internet, or if not, send back their own SMS regarding their status. This software solution is easy to implement utilizing standard Hayes commands set [24][25], commonly referred to as attention (AT) commands, to read from the mobile device. Many potential wireless devices presently have SMS capabilities build in which makes them ideal for this situation.

3.4 DATA SECURITY

Due to the sensitive nature of some information, which may be acquired through the sensor devices, such as in the case of medical monitoring [26][27] or personnel monitoring, the data must be kept secure on both the gateway and server.

Using the given framework, the gateway is fairly secure against tampering, as it does not remain on the network and even further protected against attack if Dynamic Host Configuration Protocol (DHCP) Internet Protocol (IP) address allocation is used. Furthermore, data does not remain on the gateway and is purged as necessary limiting the damage caused by a compromised system.

The server is also designed to isolate all data from each gateway. The data transmitted by the gateway is never retransmitted on the network by the server. This is to ensure that if a gateway is being imitated by another machine requesting information, no sensitive information is released.

A transmission issue lies in the transmission of data between the gateway and server, which if intercepted may reveal private information. A reliable mechanism is utilizing a Secure Sockets Layer (SSL) connection mechanism to authenticate the gateway and server [28]. This is possible as the server and gateway possess both the memory and processing power required. Utilizing a signed certificate on the server makes sure that the gateways only connect to the verified server. This ensures that all sensitive data transferred from the gateways remains secure and only retrievable by the server.

Login data pre-shared on the gateway is used to verify the identity of the gateway upon connection to both prevent unauthorized data being transmitted and to differentiate between gateways. If a gateway is compromised and its login information is obtained by an attacker, then upon detection the server can invalidate the login information and reissue the affected gateway a new login key via SMS. The gateway is also able to maintain its data in encrypted folders to increase protection against unwanted access to private data.

3.5 RELIABILITY

Since the network connections of the gateways are not reliable, and in some cases could experience a large number of drop outs, such as in the case of a mesh networked contained in a moving vehicle communicating via wireless Internet capabilities across the High Speed Data Packet Access (HSDPA) or General Packet Radio Service (GPRS) protocols, data delivery in these cases can be extremely unreliable and further care must be taken to ensure data has been transmitted when compared to a more reliable connection mechanism to guard against data loss and corruption.

To deal with this for the purposes of delivering data to the server, the gateway was designed to transmit its data files and wait for acknowledgements from the server before clearing the entry from its own data storage. This ensured that even if the connection was lost during transmission, or the server did not receive the data files the gateway would still have a copy of the data. Though this can result in redundant transmission, such as transmission errors within large files, it ensures that data is not lost.

Another reliability concern is with the update provided from the server to the gateways. Though the updates will be tested before being deployed, due to the random connectivity nature of the design, the gateways cannot be assumed to be in the same state. One gateway may have received updates to the current revision while another may not have been able to connect thus missing many of the current updates the server has been designed to detect and handle such cases through software versioning.

3.6 SUMMARY

We examined the requirements of a proposed system to obtain physiological and survey data from a real life environment in drivers for the purposes of understanding fatigue risk factors. Given the desire to conduct research with willing participants outside the laboratory, we investigated the need for a computer-based device that could be placed in their homes to collect physiological and survey data and the requirements for such device to interface with their television, allow the connection of external medical monitoring devices and the secondary storage of acquired data sets. It was also noted how such systems are also of interest in telemedicine and telecare ([3]; [4]; [5]) and that biomedical research will require such technology as it moved towards more ecologically valid research in-the-field. We also established the need for middleware that could support remotely deployed and managed devices and our choice of Java and OSGi ([6]; [7]; [8]) to satisfy these needs. We showed that our choice of a game console as a suitable hardware device over similar devices such as personal video recorders and set top boxes was due to their high availability and stable platform details.

The PlayStation-2 was shown as a potential option because it could run Linux and allowed external hardware devices to be attached and was relatively inexpensive. However, the difficulty in attempting to compile software and run it on this platform, the difficulty of finding the needed second hand hardware, its poor performance, and lack of support, all demonstrated that the PlayStation-2 would be an inferior choice of platform. We chose the PlayStation-3 as an alternative to the PlayStation-2 because it could run Linux without having to modify the console or introduce an extra "Linux kit". It had greater hardware capabilities that increased the flexibility of the type of software that could be run and how this could be developed.

It was noted that whilst the PS3 was more expensive than the PS2 (including the extra hardware and Linux kit) and that the PS3 was not expected to drop in price considerably soon, it had greater hardware availability, which may justify the extra price paid for its use in our research. At the time of writing this paper, the PS3 is our preferred computer system for the home for remote data recording and collection. We touched briefly on some network, data transmission, security, and reliability issues associated with the remote management of the devices. All are critical considerations to the efficient and cost effective deployment of the proposed biomedical data acquisition technology.

3.7 REVIEW QUESTIONS

1. What are the applications of game consoles in the area of biomedical sensing in the field?
2. What advantages are there for using game consoles over personal computers?
3. What are the security issues that need to be considered with common PC application for remote data acquisition?
4. Communication with remote systems is not always reliable. What issues need to be considered in the design of such remote systems?
5. What are some of the advantages of using the JAVA language for writing gateway device applications?
6. What are the advantages of using an OSGi platform in residential gateway devices?
7. What are the advantages of applying PS3 opposed to PS2 for building a remote gateway device?
8. Identify some applications of remote gateway systems apart from the biomedical fields.
9. List some of the ethical and privacy issues that need to be considered in the application of such remote data acquisition technology?
10. Describe a telemedicine, or smart home application of remote gateway technology.

References

[1] Lal SKL & Craig A. 2005, 'Reproducibility of the spectral components of the electroencephalogram during driver fatigue', International Journal of Psychophysiology, vol 55(2), pp137-43.
[2] Ting P et al. 2008, 'Driver fatigue and highway driving: a simulator study', Physiology and behavior, vol 94, pp 448-453.
[3] Chan, M., et al. , 'Smart homes – Current features and future perspectives', Maturitas, 2009.
[4] Chan, M.; Esteve, D.; Escriba, C. and Campo, E., 'A review of smarthomes – Present state and future challenges', Computer Methods and Programs in Biomedicine 91 (2008), pp55-81, 2008
[5] Lin, C.; Young, S. and Kuo, T., 'A remote data access architecture for home-monitoring health-care applications', ScienceDirect Medical Engineering and

Physics 29 (2007) pp199-204, 2007

[6] Li X. and Zhang W. 2004, 'The Design and Implementation of Home Network System Using OSGi Compliant Middleware', IEEE Transactions on Consumer Electronics, vol. 50, no. 2, May 2004.

[7] Zhang, H., Wang, F. and Yunfeng, A. 2005, 'An OSGi and agent based control system architecture for smart home', Proceedings of the 2005 IEEE Conference on Networking, Sensing and Control, pp13-18, Beijing, China, 2005.

[8] Kirchof, M. and Linz, S. 2005, 'Component-based development of Web-enabled eHome services', Personal and Ubiquitous Computing, vol. 9, issue 5, September 2005.

[9] Actigraph 2007, Actigraph GT1M Monitor / ActiTrainer and ActiLife Lifestyle Monitor Software User Manual, Actigraph LLC, March 2007, <http://www.theactigraph.com>.

[10] Van Hoff, A. 1997, 'The case for Java as a programming language', IEEE Internet Computing, vol. 1, issue 1, pp 51-56, Palo Alto, CA, USA.

[11] OSGi Alliance 2009, OSGi Alliance | About / The OSGi Architecture, viewed 16 March 2009, <http://www.osgi.org/About/WhatIsOSGi>.

[12] Apache 2009, Apache Felix, viewed 4 February 2009, <http://felix.apache.org/site/index.html>.

[13] Eclipse 2009, Equinox, viewed 4 February 2009, <http://www.eclipse.org/equinox/>.

[14] Knopflerfish 2008, Knopflerfish OSGi – open source OSGi service platform, view 4 February 2009, <http://www.knopflerfish.org/>.

[15] ProSyst 2009, OSGi Framework Implementations – open source Equinox and commercial – ProSyst, viewed 4 February 2009, <http://www.prosyst.com/products/osgi_framework.html>.

[16] Sony Computer Entertainment 2009, Business Development/Japan | CORPORATE INFORMATION | Sony Computer Entertainment Inc., Sony Computer Entertainment, viewed 4 February 2009, <http://www.scei.co.jp/corporate/data/bizdatajpn_e.html>.

[17] Sinclair 2006, 'Sony stops making original PS', Gamespot, viewed 4 February 2009, <http://au.gamespot.com/pages/news/story.php?sid=6146549>.

[18] Microsoft 2000, Xbox Brings "Future-Generation" Games to Life, Microsoft Corporation, viewed 4 February 2009, <http://www.microsoft.com/presspass/features/2000/03-10xbox.mspx>.

[19] Sony 2001, EE Overview, Sony Computer Entertainment Inc., version 5.0, published October 2001, Tokyo, Japan.

[20] Playstation 2 Linux Community 2007, Linux for PlayStation 2 Community: Linux for Playstation 2 FAQs, viewed 4 February 2009, <http://playstation2-linux.com/faq.php>.

[21] Kaffe 2009, Kaffe.org, viewed 4 February 2009, <http://www.kaffe.org/>.

[22] Sony Computer Entertainment 2 2008, Linux Kernel Overview, viewed 19 March 2009, <http://www.kernel.org/pub/linux/kernel/people/geoff/cell/ps3-

linux-docs/ps3-linux-docs-08.06.09/LinuxKernelOverview.html>.

[23] IBM 2004, IBM developer kits for Java technology on Apple PowerPC hardware, viewed 19 March 2009, http://www.ibm.com/developerworks/systems/library/es-apple.html

[24] HarmoniousTech Limited, Developershome. SMS Tutorial: How to Send SMS Messages from a Computer / PC? AT Commands, [Online] HarmoniousTech Limited, 2008 [Cited: March 25, 2009] http://www.developershome.com/sms/howToSendSMSFromPC.asp

[25] HarmoniousTech Limited, Developershome. SMS Tutorial: How to Receive SMS Messages Using a Computer / PC?, [Online] HarmoniousTech Limited, 2008 [Cited: March 25, 2009] http://www.developershome.com/sms/howToSendSMSFromPC.asp

[26] Jurik, A.D. and Weaver, A.C., "Remote Medical Monitoring", Computer, 2008, Vol 41, pp 96-99.

[27] Yuce, M.R., et al., "A Wireless Medical Monitoring Over a Heterogeneous Sensor Network", Engineering in Medicine and Biology Society, IEEE, 2007, pp. 5894-5898.

[28] Karlof, C.; Sastry, N. and Wagner, D., "TinySec: a link layer security architecture for wireless sensor networks" Proceedings of the 2nd international conference on Embedded networked sensor systems, ACM, 2004, pp. 162-175

Chapter 4.

An Approach to Fall Detection using Gaussian Distribution of Clustered Knowledge

Mitchell Yuwono, Steven W. Su and Bruce Moulton

Faculty of Engineering and Information Technology, University of Technology, Sydney, Ultimo, New South Wales, Australia
mitchell.yuwono@student.uts.edu.au

The increasing population of elderly people has created significant push to research in fall prevention and detection. World Health Organization noted substantial amounts of incidents and accidents among elderly people due to falls worldwide. This chapter proposes a fall detection algorithm using clustered fall signals from a single waist worn wireless tri-axial accelerometer. The method proposed is to approach fall detection using digital signal processing, data clustering, signal multiplexing, and neural networks.

4.1 INTRODUCTION

Recent demographic trend worldwide shows that there is a clear indication of an increase of ageing population. The term elderly people are generally understood as people aged 65 and above. Australian bureau of statistics reported that there are approximately 2.4 million people aged 65-84 in 2007 and is projected to grow to 6.4 million by 2056 [1]. In Japan, according to the census in 2010, there were as many as 29.29 million elderly people, which represented 24.1% of the total population which are projected to increase up to 39.6% by 2050 [2]. With such trend in elderly population increase worldwide, the increase of fall risks is quickly becoming a global health issue.

Bio-Informatic Systems, Processing and Applications, 69-82,

Falls are recognized by the World Health Organization (WHO) as a major cause of hospitalization on elderly people due to hip, limb, shoulder fractures, and fall-related traumas. Fall cases that lead to fatalities in elderly population reaches up to 36.8 per 100,000 populations and is projected to increase in an alarming rate. WHO reports that if no preventative measures are undertaken, the projected fatalities associated with falls will double over 20 years period [3].

Current research in University of Technology, Sydney uses Ambulatory accelerometer devices to detect falls in a controlled environment [4]. Acceleration measurements are desired to indicate abnormal body activities such as tremors and falls in older people. In this chapter we will discuss an approach to automatic fall detection using signal processing, clustering and neural network technique from a waist-worn tri-axial accelerometer.

4.2 OVERVIEW

The proposed fall detection method can be seen in the block diagram in Figure 4.1.

The data flow block diagram is divided into two important stages: online stage and offline stage. The online stage includes data collection and classification was done under Java 2 Standard Edition. The offline stage includes data clustering and neural network training was done under MATLAB®.

The first stage is data collection. In the data collection stage the sampled raw acceleration signal in discrete time was captured using a tri-axial accelerometer. These signals are used to construct the training data for the offline clustering and

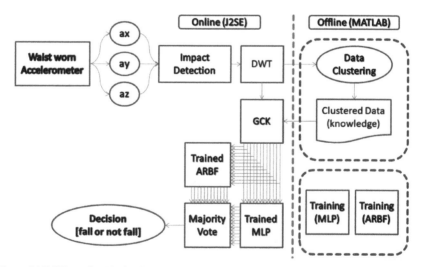

Figure 4.1 Fall Detection Block Diagram.

learning stage. From the collected data, training data is generated by observing the time domain trend of the sum of squared acceleration ($g^2 = a_x^2 + a_y^2 + a_z^2$).

Impact detection is initiated by using a primitive acceleration threshold method set at a low level ($3g^2$). When an impact is detected the data is captured and processed. The visualization window of the process can be seen in the Figure 4.2. These preprocessed data are used to do the clustering and learning.

After the offline clustering and learning stage, the knowledge database and trained neural network are implemented in the online classifier and the online classification stage will be ready to be carried. In this stage, impact detection will be done in real time as the raw signals are streamed in the input buffer. When acceleration signal with g^2 higher than 3 is detected, the signal is captured and pushed to the classification queue. The classification stage is divided into three sub-stages: Gaussian distribution of Clustered Knowledge (GCK) signal generation, Neural Network classification, and Decision Making using majority vote.

4.3 SIGNAL PREPROCESSING

Basic information regarding falls can be obtained by examining acceleration signal using waist-worn accelerometer. One of the more popular algorithms in fall detection is threshold-posture method. A fall is when a person experiences

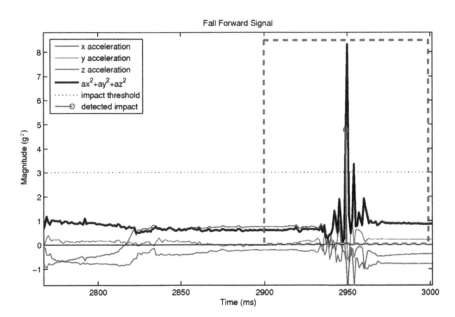

Figure 4.2 Primitive Thresholding method.

acceleration above a certain threshold – that is, the impact acceleration as a person hits the ground – followed by a change in body posture and a period of lying down/not moving [4]. A typical fall signal can be seen in Figure 4.4. Using Discrete Wavelet Transform (DWT) to extract the important features, we reduce this signal dimensionality from 100 to 14. The Discrete Wavelet transform can be described simply as a series of cascaded low pass filters ($G(z)$) and down-sampling ($\downarrow 2$) blocks. This can be seen in Figure 4.3. We have observed that using DWT to the 3rd order sufficiently extract the important features of the time domain signal. Figure 4.5 presents extraction result.

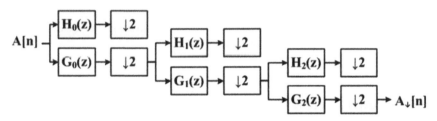

Figure 4.3 3rd order Discrete Wavelet Transform.

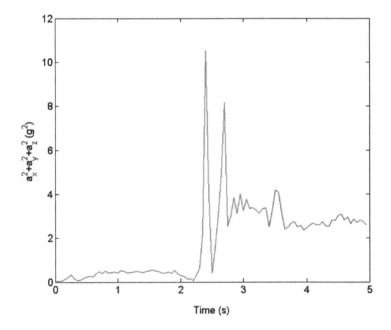

Figure 4.4 Fall signal characteristics.

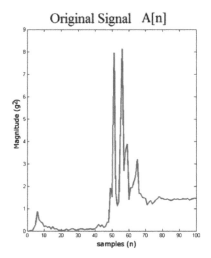
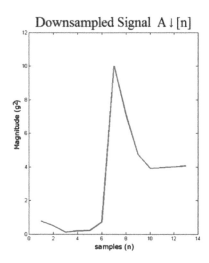

Figure 4.5 Feature Extraction using Discrete Wavelet Transform.

4.4 DATA CLUSTERING

Data clustering is an important step to group signals together according to its degree of similarity. The input of this step is the preprocessed data. The output of this step is the clustered data (knowledge). Clustering can be defined as grouping similar signals together by measure of similarity. In our application the squared Euclidean distance is used as the measure. The process of clustering can be done using any unsupervised learning algorithm such as K-means, Fuzzy C-Means, Self Organizing Map, Swarm Algorithms, etc. In the previous experiment we were using the Regrouping Particle Swarm Algorithm modified for clustering application [4]. The clustering result is illustrated in Figure 4.6. Figure 4.6 presents a projection of the 13 dimensional data on a 2-Dimensional plane using Manilla's Random Projection [5]. Colored dots are clustered datapoints, blue x signs are estimated centroid locations. It can be seen in Figure 4.6 that Activity of Daily Living (ADL) data is grouped in the middle 0, while abnormal activities data are scattered over the projected dimensions.

Clustered data is stored inside a database we refer to as "knowledge" database: a compressed representation of the whole dataset. The storage method is simply by calculating mean and standard deviation of each cluster. An illustrative example is shown in Figure 4.7 and Figure 4.8. In Figure 4.7 150 individual preprocessed fall signals which are member of a cluster are stacked for visualization purpose. The sample number indicates signal number 1 to 150. Figure 4.8 illustrates how a cluster is stored inside the knowledge library. The vertical lines in each dimension are the standard deviations.

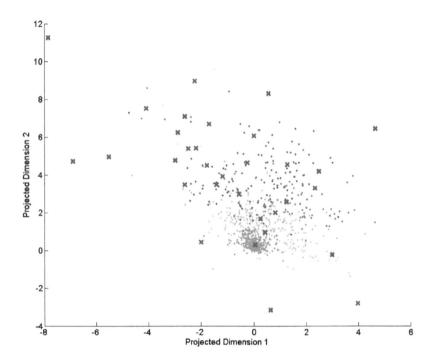

Figure 4.6 Datapoints and Cluster Centroids projected on a 2-Dimensional space.

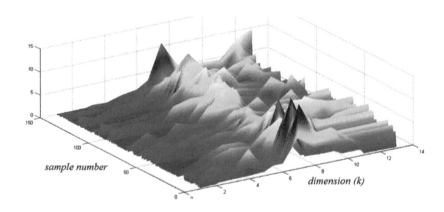

Figure 4.7 Visualization of 150 individual fall signals in a cluster.

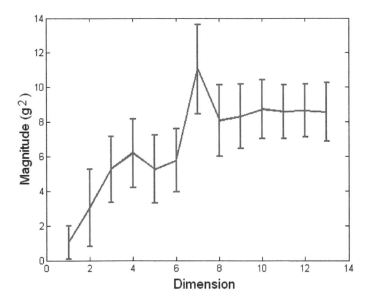

Figure 4.8 Mean and standard deviation of each dimension of a cluster.

4.5 GAUSSIAN DISTRIBUTION OF CLUSTERED KNOWLEDGE

Gaussian distribution of Clustered Knowledge (GCK) is a signal multiplexing method inspired by Monte Carlo method. The GCK method is intended to help classify clustered patterns of statistical characteristics. It refers incoming input signals to cluster centroids, and multiplexes them based on the Gaussian characteristics of the clusters. Each input signal is queried against the cluster centroids and passed through a Radial Basis Kernel to get the rate of membership. The cluster with the highest rate of membership is selected as the GCK seed. Similar result can be achieved by selecting the cluster with the least Euclidean distance to the incoming signal as the GCK seed. The seed selection scheme can be seen in Figure 4.9.

A knowledge signal is obtained by generating a vector of Gaussian random numbers based on the cluster's mean and standard deviation as stored in the knowledge database. The generated signal is fused with using a significance ratio of A:B to create a GCK signal. The block diagram can be seen in Figure 4.10.

4.6 NEURAL NETWORK

The extracted data are then used to train two different types of classifier. The first type is a standard Multilayer Perceptron (MLP) type, the second type is a

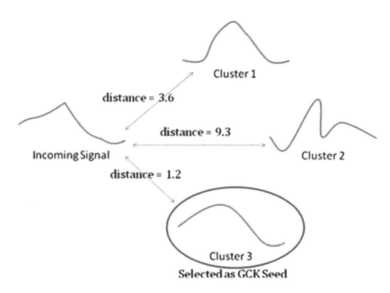

Figure 4.9 Seed selection scheme.

probabilistic neural network we call the Augmented Radial Basis Neural Network (ARBF). MLP and ARBF are chosen because of their different characteristics [4,6]. ARBF have previously been used in time signal classification of head movement patterns, with promising results [6]. ARBF consists of an RBF layer

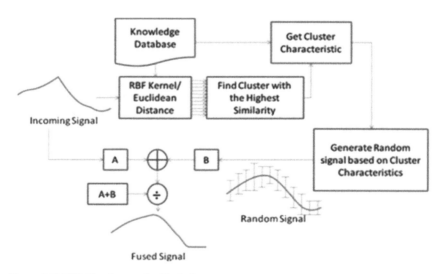

Figure 4.10 GCK Signal generation block diagram.

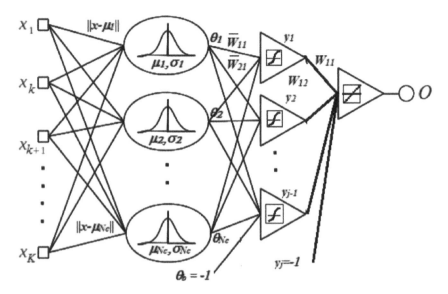

Figure 4.11 ARBF Neural Network Configuration.

and an MLP augmentation layer, shown in Figure 4.11. ARBF is reported to have a sensitivity advantage over conventional RBF and a specificity advantage over MLP [6]. The RBF centroids of the radial basis layer are the centroids obtained from the clustering process. This scheme allows the network to improve as better knowledge database is obtained. Note that x is the incoming signal, k is the signal dimension. The standard deviation of every basis should be manually tuned to best training result. σ is variable according to the number of cluster and the standard deviation of the cluster. Best result is usually achieved by tuning σ to the standard deviation of the cluster added with a constant bias value.

The Neural Networks are trained using the Neural Network Toolbox available in MATLAB® summary of training steps is as follows:

1. MLP is trained using resilient backpropagation.
2. Train ARBF:
   a. Create an RBF layer using the cluster centroids taken from the knowledge database as a result of the clustering process
   b. Pass the pre-processed data to the RBF layer and take the output of the layer as the input to the MLP layer
   c. Train the MLP layer with resilient backpropagation
   d. Merge the RBF and MLP layer together.

After the training, the two networks are used together in parallel in the final application. This way we are creating an ensemble of two different types of neural networks. The final decision of the ensemble is chosen by using majority voting. This scheme can be seen in Figure 4.12.

4.7 DATA COLLECTION

A data set referred to as the In-group fall data was collected from 5 healthy volunteers, 2 females and 3 males. This was made up of 293 fall signals, 153 signals used for training, and 140 signals used for testing (in-group performance). Out-group fall test data (used for testing) was collected from 3 different healthy male volunteers whose data was not included in the training data. This set included 85 signals, all used to test "out-group" performance. out-group meaning that these people's data was not used as training data. All of the volunteers were aged between 19 and 28 years.

A total of 293 fall signals were recorded. 153 signals were used for training, 140 were used to test the system for in-group performance, 85 fall signals were recorded to test out-group performance. The Activities of Daily Living (ADL) training data was collected from 3 people. A total of 8 hours of ADL data was collected in a home environment. An additional hour of exercise data was recorded from 2 people in a gym environment. A total of 1831 ADL signals were collected. 1000 randomly selected ADL signals were used for the training set while 831 were used for testing. Of the 1000 randomly selected signals used for training, 750 related to ADL routine, and 250 related to ADL exercise. Of the 831 signals used for testing, 400 related to ADL routine and 381 related to ADL gym exercise. A validation set was randomly taken from the training set with the ratio of training versus validation = 4:1.

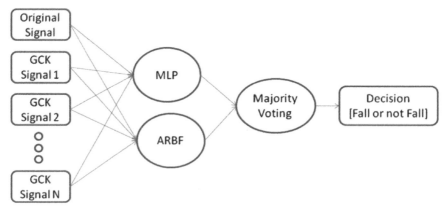

Figure 4.12 Classification scheme.

4.8 RESULTS AND DISCUSSION

Experimental results can be seen in table 4.1. Trials indicated that sensitivity improves up to GCK = 5 but stays the same at greater than 5, while specificity of the system decreases when GCK greater than 5 is used. Using 5 GCK signals, the approach achieves 100% sensitivity on in-group falls, 97.65% on out-group falls, 99.33% specificity on routine ADL, and 96.59% specificity on exercise ADL.

There are limitations applicable to these results. First limitation is posed by the relatively small number of subjects. Second is because the training data consists only of controlled falls where subjects fall deliberately on a mattress – not accidental falls as is happening in the real world situation. Third is because the ADL data included only home and gym activities – data relating to ordinary work, transportation and other non-home activities would likely be more representative of some people's typical daily activities which is missing in this preliminary stage. Fourth is the data was acquired from people aged 19-28 years – it would be preferable for future work to include people from older age brackets.

Table 4.1

Experimental result of different classifier schemes with variable number of GCK signals

GCK Signals (N)	Classifier	Ingroup Falls	Outgroup Falls	ADL (routine)	ADL (exercise)
0	MLP	96.43%	89.29%	99.33%	96.85%
	ARBF	92.59%	85.88%	100.00%	98.95%
	MLP+ARBF	96.43%	92.94%	100.00%	98.95%
5	MLP	97.14%	95.29%	99.33%	95.28%
	ARBF	95.56%	92.94%	99.78%	96.06%
	MLP+ARBF	98.57%	97.65%	99.56%	96.85%
10	MLP	100.00%	95.29%	99.33%	95.28%
	ARBF	96.30%	88.24%	99.78%	97.11%
	MLP+ARBF	100.00%	97.65%	99.33%	96.59%

4.9 CONCLUSIONS

This chapter has provided a general overview of the newly proposed method on fall detection. The proposed methods for signal preprocessing, data clustering, GCK signal multiplexing, and ensemble neural network model for the particular application has been explained. The chapter also demonstrates the effect of the method on the experimental data. From experimental results we conclude that adding GCK signals improves sensitivity in expense of specificity. We also conclude that a more reliable classifier can be made using ensemble of MLP and ARBF.

These results are an exploratory step towards gaining knowledge about potential elements of a fall detection system. The implications of the results are somewhat limited due to limitations of the data acquisition processes. The method described in this chapter thus needs further development and experimental investigation.

4.10 PROBLEMS AND SOLUTIONS

This chapter offers a solution to problem of accelerometry fall signal processing and recognition.

References

[1] Australian Bureau of Statistics, "Future population growth and ageing", available at http://www.abs.gov.au/AUSSTATS/abs@.nsf/Lookup/4102.0Main+Features 10March%202009

[2] Japan Statistics Bureau, "Statistical Handbook of Japan 2011 – Chapter 2: Population", available at http://www.stat.go.jp/english/data/handbook/c02cont.htm

[3] World Health Organization, "WHO Global Report on Falls Prevention in Older Age", France, 2007.

[4] M. Yuwono, S.W. Su, B. Moulton, "Fall Detection using a Gaussian Distribution of Clustered Knowledge, Augmented Radial Basis Neural-Network, and Multilayer Perceptron", in Proc International Conference on Broadband Communications & Biomedical Applications (IB2COM), November 21-24, pp. 145-150, 2011.

[5] H. Mannila, E. Bingham, "Random projection in dimensionality reduction: applications to image and text data", in Proc of the ACM SIGKDD International Conference on Knowledge Discovery and Data Mining, pp. 245-250, 2001.

[6] M. Yuwono, A.M.A Handojoseno, H.T. Nguyen, "Optimization of head

movement recognition using Augmented Radial Basis Function Neural Network", in Proc Annual International Conference of the IEEE EMBS, August 30 – September 3, pp. 2776-2779, 2011.

Chapter 5.

ZigBee Sensor Network propagation analysis for health-care application

Roberto Maurizio Pellegrini[1], Samuela Persia[2], Diego Volponi[1] and Giuseppe Marcone[1]

[1]*Fondazione Ugo Bordoni,* [2]*Fondazione Ugo Bordoni & University of Rome "Tor Vergata"*
pellegrini@fub.it, spersia@fub.it, dvolponi@fub.it, gmarcone@fub.it

5.1 HEALTH CARE SCENARIOS

The goal of this chapter is to investigate a possible application of Wireless Sensor Networks for healthcare. In particular the scope of this study is to design a "Smart Hospital" and thus to define main performance requirements for a ZigBee Sensor Network deployed in real hospital environment. It can be possible by considering the RF propagation of ZigBee devices operating within a hospital environment and after by analyzing their real indoor behavior to design healthcare systems.

WSNs have attracted a growing interest in recent years thank to their versatility of being employed in different application fields [1]. In the last two decades extensive researches have led to the definition of new generations of wireless systems able to integrate communication, localization and/or location tracking services to improve accuracy and reliability and enlarge potential application fields. Specifically indoor positioning and location tracking systems are considered an excellent way to improve productivity and optimise resource usage when applied to a wide range of scenarios: industrial, healthcare, commercial and military as well.

The most common performance questions for wireless communications involve range. The RF propagation in indoor environment is hard to predict, due to the extensive interference and thus a consistent propagation model is difficult to be derived.

Bio-Informatic Systems, Processing and Applications, 83-106,

In this chapter we proposed a RSSI measurement process in a specific indoor environment to characterize the main ZigBee propagation features.

In particular we firstly carried out different measurements as case studies and we also derived a propagation model to fit with experiments.

On the basis of the experimental results it has been possible to analyze the different factors that affect the measurements, such as external factors (e.g. multipath, fading) or internal ones (e.g. hardware device, integrated antennas). For this scope several studies have been proposed in literature. In [3] authors investigated the external factors effect on the RF characteristic. The RF activity for the 24 hours period.

On the other hand some works have analyzed the effects of the internal factors over the RSSI measurements, such as the effect of the antenna polarization between the transmitter and the receiver device [4], or the effect of the hardware design of the devices [5]. In this way we try to isolate the different factors (external and internal) that influence the RSSI measurements in order to define a propagation model that well fits with the experiment results.

Furthermore to deeply define the ZigBee behavior we have compared the measurements with a suitable propagation model.

This preliminary study permitted us to include the main measurements outcomes in a planning simulator which emulates the radio propagation in hospital environments to define in the future the infrastructure requirements for a "Smart Hospital" definition. We remind that is not allowed to perform experiments in hospital enviroment and then it is important to define a suitable propagation-planning simulator. This analysis has the aim to verify the ZigBee propagation features in a specific indoor context and thus to derive accurate models useful for next generation of healthcare purposes and applications.

5.2 HEALTH CARE SCENARIOS

This section provides a brief overview of the healthcare scenarios especially in which the deployment of Wireless Sensor Networks (WSNs) represents an useful improvements opening a door in different applications that never existed in the past. For instance, healthcare applications have growing interest due to the possibility to reduce healt-care expenses.

Generally, healthcare scenarios can be mainly divided into two categories:

- In-Hospital Scenarios - in order to design "Smart Hospitals"-
- Outside-Hopsital scenarios - which are the basis on the "Telemedicine".

In the following a brief overview of them is showed. For the interested reader a more deep description is available in [6].

(a) In-Hospital Scenarios

       – Medical Equipment Localization and Tracking.
       – Tracking Vital and Highly Sensitive Medical.
  (b) Outside- Hospital Scenarios
       – Mobile Telemedicine Services;
       – "Smart-Home" for Elder Health-Care

5.3 ZIGBEE CHARACTERISTICS

ZigBee refers to a suite of high level communication protocols using small, low power devices based on IEEE 802.15.4 standard for wireless personal area networks [7]. This technology is simpler and cheaper than any other WPANs (see Figure 5.1).

ZigBee is targeted at Radio Frequency applications that require a low data rate, low power usage, and secure networking. 802.15.4 defines the physical layer, and ZigBee defines the network and application layers. In other words ZigBee is a low cost, low power , wireless mesh networking standard. The low cost allows the technology to be widely accepted and deployed in wireless control and monitoring applications, the low power usage allows longer life with smaller batteries and mesh networking provides high reliability and larger range.

The ZigBee standard is based on IEEE 802.15.4 physical radio specification and operates in unlicensed bands worldwide at the following frequencies: 2.400 2.484 *GHz* , 902-928 *MHz* and 868.0 - 868.6 *MHz* . ZigBee uses the same channel set as specified in 802.15.4. In the 2.4 GHz band, these channels are numbered 11 through 26. Channel numbers 0 through 10 are defined by the sub-1 GHz 802.15.4 radios, but ZigBee (at least to date), doesn't run on the sub-1 GHz radios as shown in Figure 5.2.

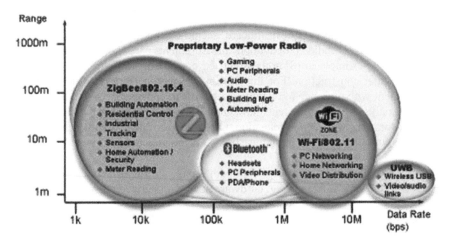

Figure 5.1 Low rate RF devices [2].

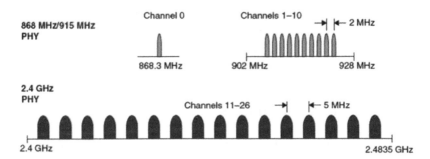

Figure 5.2 Zigbee used frequencies [2].

The standard suggests that the ZigBee-based WSN can be composed by three different types of devices, such as:

- *ZigBee Coordinator* (ZC), or the the sink node of the network, responsible of network set up, data collecting, and network management;
- *ZigBee Router* (ZR), or the Full Function Device (FFD) of the network, responsible of routing function;
- *ZigBee End Device* (ZED), or the Reduced Function Device (RFD) of the network, responsible of detecting events, and trasmitting data towards devices (ZC or ZR) within its coverage range.

For our purposes we consider a simple WSN composed by a ZC device (the transmitter) and a ZR device (the receiver).

5.4 PROPAGATION ANALYSYS IN HOSPITAL ENVIRONMENT

To define requirements for healthcare scenario, the propagation analysis is needed. Actually, radio propagation measurements are not possible within hospitals. For example in [8] authors carried out propagation measurements in a hospital under construction. For this reason, we performed measurements in an environment with presence of metallic cupboards to consider a real environment.

By analysis of measurements outcomes it is possible to define which are the main factors that influence propagation. On the basis of the experimental results it has been possible to observe that measurements are affected by:

- External factor - e.g. multipath, fading
- Internal factor - e.g. hardware device, integrated antennas

To derive the ZigBee propagation features we fit measurements with a reasonable model in which the main factors (both internal and external), responsible of indoor propagation characterization, have been included. This study permitted us to include the main measurements outcomes in a simulator (explained in the following section) which emulates the radio propagation in hospital environments.

5.4.1 Experimental set up

For our experiments, we used low-power RF serial modules ZB01CA produced by Digi (see Figure 5.1). All modules support the ZigBee stack: wireless RF, PHY layer, and MAC layer are integrated in ZB01CA. For our experiments we used two ZigBee sensors mounted on ZB01CA modules, one set as ZC and the other as ZR.

The ZB01CA modules were connected to the PC via RS232 port, and the ZigBee sensors configured and controlled by a specific application.

The main characteristics of the ZigBee modules used are depicted in the Table 5.1. The transmitting and the receiving nodes were both power supplied by a $3V$ block battery.

Measurements were carried-out inside a room with some fixed obstacles e.g. metallic cupboard and reinforced concrete walls (see the planimetry of the room in Figure 5.3). We used one fixed-node connected to a PC for data gathering and another mobile-node. Both nodes were placed at 1 m from the floor,

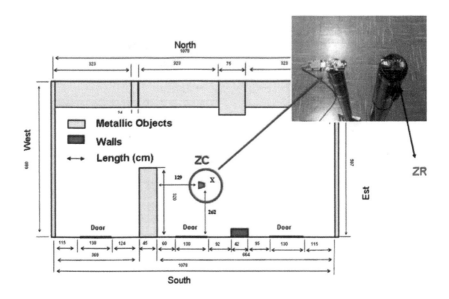

Figure 5.3 Indoor environment planimetry and ZB01CA devices (ZC and ZR) on the upper side.

Table 5.1

Main ZigBee Characteristics

Parameter	Symbol	Value
Modulation	DSSS - QPSK	2 bit/symbol
Operational Band	BW	ISM (2.4 GHz)
Bit Rate	R_b	250 kbps
Transmitted Power	P_{TX}	17 dBm
Receiver Sensitivity	S	-102 dBm
Indoor-Urban Range	R	100 m

thus we have evaluated the RF propagation of the ZigBee devices by measurements on a plane. In the ideal conditions, a simple Omni-directional ZigBee antenna would produce a linearly polarized radiation, and then the maximum of the power transferring between two antennas is achieved when they have the same polarization. However in real conditions the antenna polarization is

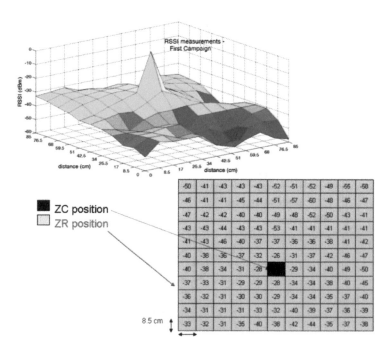

Figure 5.4 *RSSI* vs. distance in the first campaign. Measurements were carried out with transmitter in a fixed location (black square in the lower figure). *RSSI* was measured for different positions around the transmitter with step of 8.5 *cm*. The direction of the transmitted and the received antenna is the same for all measurements.

not linear and not trivial to predict. Therefore reciprocal orientation of the antennas has to be empirically considered to evaluate the maximum transferring power. Several studies have been proposed about this topic, and it has been demonstrated that also in real conditions the highest *RSSI* values are obtained when the polarizations of transmitter and receiver are the same [4]. For these reasons we placed the ZigBee modules parallel and we performed the measurement campaigns. In general the measurements have been carried out as follows: the ZC sent packets to the receiver, and the router measured the *RSSI* of the received packet. We extracted *RSSI* data byte from the received packet by a query of the Coordinator which was connected to the PC via RS232 port. As already mentioned to isolate different factors we performed:

- **First Campaign** - We carried out measurements for each position nearby the fixed node, i.e. in front, behind, on the left, and on the right of ZC, and thus not solely along a direction such as in [4]. Specifically measurements were made within a square area of side equal to $93.5 \, cm$. The ZC device was placed in the center of the area and the ZR was moved in different positions (about 120) in all directions around the ZC on a grid of $8.5 \, cm$ step (See Figure 5.2).

  The ZC transmitted power for this campaign was $0 \, dBm$ in order to not saturate the ZR receiver front-end.

- **Second Campaign** - To extend the analysis performed during the first campaign, in order to minimize the impact of the different effects, e.g. multipath, fading, and antennas polarization, further measurements in a restricted rectangular grid $34 \, cm \times 25.5 \, cm$ were made. ZC was in a fixed location at about $10 \, m$ from the ZR, that was moved in different positions with step of $8.5 \, cm$. In this way being the receiver far from the transmitter ZC (See Figure 5.3), we have assumed the impact of the mentioned effects as constant in that small area, thus we have evaluated the RF propagation behaviour isolating the influence of the various factors. The ZC was in a fixed position with a transmitted power of $17 \, dBm$.

By this analysis we can define the RF indoor propagation in different scenarios.

5.4.2 Experimental results

The RSSI vs the distance measurements have been collected for different campaigns. In particular:

- **First campaign** - These measurements were performed to identify the best position that permits the maximum match between the two antennas. In case of misalignment of the polarization between the two antennas a loss factor has to be taken into account. As mentioned in previous sections in real conditions, the antennas produce a field with polarization in more than one direction, and in addition the presence of objects in indoor environment leads to received signals not aligned with the polarization of the received antenna. The best polarization angle between the two antennas, which guarantees the minimum loss, has to be empirically calculated on the basis of measurements evaluations. Hence considering the grid of measurements depicted in Figure 5.4, we can see that the maximum values are reached along the two diagonals of the square centered on the ZC module.

- **Second campaign** - Once investigated the positions of the two devices to maximize the *RSSI* values, further measurements were carried out to estimate the RF propagation behaviour of the transmitted signal in real condition, under the assumption that the propagation factors are constant within the monitored area. The results confirmed our assumptions, and the RF indoor propagation presents a linear decay with respect to the distance with weak non-linear effect (see Figure 5.5).

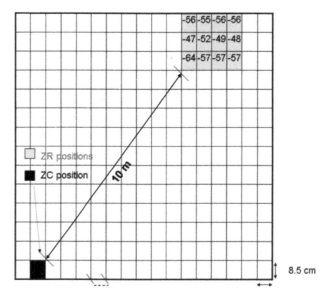

Figure 5.5 *RSSI* vs. distance in the third campaign. Measurements were carried out with transmitter in a fixed location (black square in the lower figure). *RSSI* was measured for different positions 10 *m* far from the transmitter with step of 8.5*cm*. The direction of the transmitted and the received antenna is the same for all measurements.

5.4.3   Propagation Model

The effectiveness of a propagation model depends on how the theoretical approximation can fit with the measurements. For our purpose the following model for $RSSI$ (dBm) has been used:

$$RSSI = P_{Tx} - P_{LOS} \tag{5.1}$$

where the P_{Tx} is the transmitted power, and P_{LOS} is the propagation loss expressed by the log-distance path loss model that has been used extensively in literature:

$$P_{LOS} = PL(d_0) + 10 \cdot n \log_{10}\left(\frac{d}{d_0}\right) \tag{5.2}$$

where n is a path loss exponent which indicates the decreasing rate of signal strength in an environment, d_0 is a reference distance which is close to the transmitter, and d is the distance between the transmitter and receiver. In general, the exponent n is environment-dependent, in a free space its value is close to 2. $PL(d_0)$ can be obtained by measurements. Actually the model expressed by (5.2) does not take into account the factors in the surrounding environment, such as interference, multipath, and fading. This means that, the $RSSI$ values measured not corresponding to the empirical values evaluated by the model. In addition, remembering that the received signal - the vector sum of the direct wave plus all the reflected, scattered, diffracted, and/or refracted waves - is position and time dependent, and thus, in a particular location, the average $RSSI$ will also fluctuate with time. Hence the values of $PL(d_0)$ and n could not be considered as constants. Furthermore also internal factors, as e.g. the antennas polarizations, which affects the outcome of RSSI value, have to be taken into account.

As a result the RSSI at any distance d can be considered dependent on both the average path loss value (where the $PL(d_0)$, and n take into account the indoor fluctuant propagation effects) and loss due to the polarization of the antennas.

Hence:

$$RSSI = K - 10 \cdot n \log_{10}\left(\frac{d}{d_0}\right) \tag{5.3}$$

where K is the *"average loss"* factor due to both environment and polarization fluctuant effects. As mentioned in the previous section the mismatch loss due to the different polarization of the two antennas, corresponds to a "depointing" loss, i.e. the misalignment between the polarizations of the transmitting and receiving antennas.

Indeed K is the "average loss" factor due to both environment and polarization fluctuant effects.

The introduction of this parameter in (5.3) is due to the following considerations. In the ideal conditions, a simple Omni-directional ZigBee the maximum of the power transferring between two antennas is achieved when they have the same polarization.

However in real conditions the antenna polarization is not linear and not trivial to predict. In fact most antennas used in low cost short range applications do produce a radiation with polarization in more than one direction. Furthermore the environment can affect the polarization due to the multipath: if the objects that reflect the signals are not aligned or parallel with the polarization of the incident signal, the reflected signal will undergo a polarization shift [14], which could be also time dependent (see Figure 5.6).

Therefore reciprocal orientation of the antennas has to be empirically considered to evaluate the maximum transferring power [15]. Our experiments have been proposed to investigate about this topic: the role of the polarization in the maximum transferring power definition.

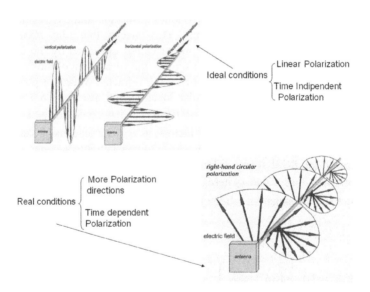

Figure 5.6 Linearly Polarized Radiation (on the left), not Linearly and Time dependent Polarization (on the right). [16]

This effect is taken into account by introducing an additional factor in the path loss model, the "*mismatch loss*" factor indicated with α_{pol}. Actually α_{pol} cannot be considered neither linear nor time independent and thus it is not possible to evaluate it analytically; thus we estimated it empirically. By measurement observations we found that the minimum loss is reached along the diagonals of the squared area monitored (i.e. $\alpha_{pol}\big|_{\theta=45^0} \approx 0 = 45$), and thus we can evaluate an average value of the α_{pol} for the different angles between the two antennas:

$$K\big|_\theta = RSSI(d_0)\big|_\theta - \alpha_{pol}\big|_\theta \qquad (5.4)$$

where for example $RSSI(d_0)\big|_{\theta=45^0} = -28dBm$ is the average value of the *RSSI* along a diagonal, i.e. for the minimum mismatch, for the first measurement campaign. By it and by considering the average value of *RSSI* for the other angles at the same reference distance $d_0 = 8.5cm$ we can evaluate the loss due to the different angles. This procedure has been applied also for the other experiments performed.

For our analysis we considered for the first experiment $d_0 = 8.5cm$, and for the second $d_0 = 1m$. For what n concerns we estimated different values for the two campaigns (see the following Section for details). Hence we compared the figures obtained by the model of (5.3) with the measurements performed in the different scenarios. Hence we considered α_{pol} depending on the angle between the horizontal axis passing for ZC and the conjunction between transmitter and receiver.

5.4.4   Analysis results

The comparison between the propagation model and the measurements has been carried out for both measurement campaigns. Specifically we observed different behaviour for measurements done in front of or behind the transmitter, due to the presence of different objects in the laboratory. Hence to fit the experiment results with the model, we set different values for the parameters in the model (5.3) and in the model (5.4). We assumed for the front *RSSI* estimation $n = 2.5$ and $\alpha_{pol}\big|_{\theta=0^0,45^0,90^0,135^0} = 0.3,0,1.4,1.2 \quad dB$, while for the *RSSI* estimation behind the ZC $n = 1.6$ and $\alpha_{pol}\big|_{\theta=180^0,225^0,270^0,315^0} = 3,0.4,4,3.8 \quad dB$ respectively for the first campaign. In the second campaign, only a polarization angle was considered, and for it we assumed $\alpha_{pol}\big|_{\theta=45^0} = 0.7 \quad dB$, and $n = 1.6$. In the following, the main evaluations of the comparison are reported.

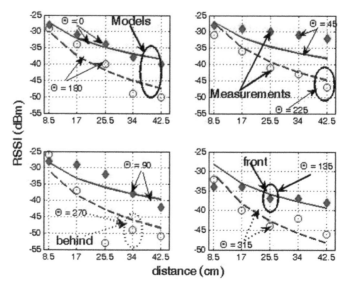

Figure 5.7 *RSSI* vs. distance. Comparison between the first campaign measurements and the propagation model. The model is characterized by $n = 1.6$, and $n = 2.5$ for front and behind measurements comparison respectively. Different polarization angles are considered (up-left: $\theta = 0,180^{0}$; up-right $\theta = 45^{0},225^{0}$; down-left $\theta = 90^{0},270^{0}$; down-right $\theta = 135^{0},315^{0}$.

- **First campaign analysis** - The rate of *RSSI* decays over distance is higher for the positions of the receiver in front of the transmitter, rather than the positions located behind it.

  By comparison, we noted that there is a fairly difference between the measurements and the propagation model trend for $\theta = 45^{0}$ and $\theta = 225^{0}$ with respect to the other angles: it means that for these angles the measurements experiment a linear trend while for the other angles they experiment fluctuant trends. This is because the values of K and n are both affected by environmental factors and are time dependent: they vary significantly between different locations and then a linear model does not fit well. On these axes there is the best polarization angle between the antennas and thus we could suppose in real conditions a bipolar behaviour of the ZigBee modules antennas (see Figure 5.7).

- **Second campaign analysis** - Only a polarization angle $\theta = 45^{0}$ was taken into account. We considered the model with different values for loss due the polarization and loss due to the environment. We observed that the measurements fit with the model when the polarization effect is negligible and the multipath can be considered constant. We observed that the *RSSI* decays with the distance with a linear trend as expected (see Figure 5.8).

5.5 INFRASTRUCTURE REQUIREMENTS FOR AN INDOOR HELATH-CARE SCENARIOS

We considered a particular scenario to define a *'Smart-Hospital'*: Medical Equipment Localization and Tracking as depicted in Figure 5.9. The reliability of the system strictly depends on the amount of the ZigBee transmitters needed to obtain a suitable localization, and then their coverage range to cover the whole interested area. It is essential to know how many of these receivers are needed on the field since indoor locations present very severe attenuation factors varying from place to place. To define infrastructure requirements we considered measurement results obtained in the previous sections and included them in a proper simulator. This study is intended to identify a set of key parameters and variables which define the infrastructure requirements for an indoor asset tracking system applied to a generic hospital case and to analyze the effect of their variations on the coverage area and consequently, on the expected number and the location of receivers to be placed. For this scope a simulation approach has been proposed (see the following sections for the details), in order to define the number of base stations to be placed in the area.

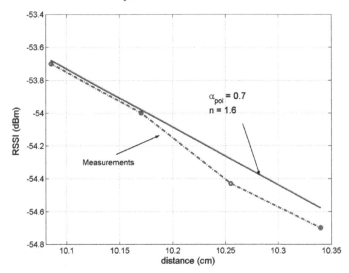

Figure 5.8 *RSSI* vs. distance. Comparison between the second campaign with the propagation model. Figure shows the model for polarization angle between transferring and receiving electromagnetic fields equal to $\theta = 45^0$ with $\alpha_{pol} = 0.7$ and $n = 1.6$.

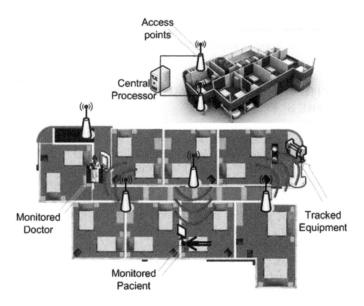

Figure 5.9 '*Smart-Hospital*' scenario [12].

To evaluate the minimum number of nodes needed for localization purposes we have to remind that in typical positioning systems, a number of BSs must detect the signal from a tag simultaneously to allow its position to be calculated. The number of base stations involved depends on the architecture of the system and whether a 2D or 3D position is to be computed.

For the purpose of this analysis, a TDOA (Time Difference of Arrival) 3D positioning system architecture is assumed. According to [13] a set of four base stations represents the minimum required system infrastructure (a cell) to allow a tag's 3D-position to be found.

Therefore the total number of nodes needed to localize object in the whole indoor area is obtained by the following equation:

$$N = 4 \left\lceil \frac{A_T}{A_c} \right\rceil \tag{5.5}$$

where A_T is the total required coverage area, and A_C is the area covered by a single tag, which depends on the tag-BS link range, r, that can be estimated according to the proper link budget. Specifically the tag-BS link range may be affected by several factors, including:

- Maximum permissible tag signal (PSD);

- Operating frequency;
- Receiver efficiency;
- Environmental factors causing attenuation such as number of walls, floors, construction materials, people.
- Overlapping between adjacent receiving stations
- Multipath

There are other factors which make propagation in hospitals even more challenging. The spotty coverage is a common issue. There is lots of metal in medical facilities that causes multipath propagation and signal attenuation. Almost every hospital uses stainless steel carts to roll around surgical equipment and other patient related materials. These carts are moving obstacles to the receivers, causing sporadic holes in radio frequency coverage.

In conclusion the scope of this study is intended to identify a set of key parameters and variables which define the infrastructure requirements for an indoor asset tracking system applied to a generic hospital case and to analyze the effect of their variations on the coverage area and consequently, on the expected number of receivers and the best locations where these receivers have to be placed.

Hence the reliability of the proposed system strictly depends on the estimation of the coverage cell area A_C of each node, taking into account all influencing propagation factors. For this scope a more details propagation model in the hospital is needed. The details are provided in the following section.

5.5.1   Propagation Model in the Hospital

For heterogeneous propagation conditions, a model can be adopted able to account for losses introduced by a number of walls: the Multi-Wall Path Loss Model. This model is largely suggested in literature [11] and it separates the calculation of these kinds of losses from the attenuation due to the link distance, and it can be expressed as:

$$PL_{M_w(d)} = PL(d) + M_w(d) \tag{5.6}$$

where M_w is the *multi-wall component* and the term $PL(d)$, accounting for the attenuation due to the link distance, as expressed in (16.2) with exponent $n = 2$. Actually the equation may be rewritten considering the antennas loss factor as described in the experimental results session and thus we have:

$$PL_{M_w(d)} = PL(d) + M_w(d) + \alpha_{pol} \tag{5.7}$$

and the multi-wall component can be considered as:

$$M_w = \sum_{i=1}^{l} k_{wi} l_i \tag{5.8}$$

being k_{wi} the number of penetrated walls of type "i" and "l_i" the attenuation due to the wall of type "i". Specific walls attenuation are considered, according to values reported in [8], where the attenuation between adjacent rooms is extracted by averaging a number of measurements, in order to remove all effects related to local characteristics of the walls. Actually, reflections as well as transmissions through slots and openings in the metallic surface occur, that can markedly influence the propagation along each path and this effect is even more evident in rooms with metallic wall elements (MRT, X-ray and operating rooms).

Therefore an average estimation of such effects must be taken into account, the walls attenuations are extracted. In particular, losses due to each kind of wall in the area, l_i, are determined by considering the corresponding attenuation increases with respect to the free space loss, at the operating central frequency $f_c = 2.4\ GHz$. These wall attenuation values l_i are reported in the Figure 5.10 with the corresponding simplified scheme of the hospital floor taken into account as a reference in our studies.

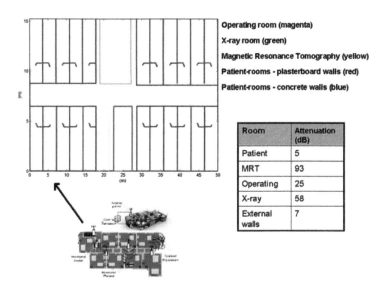

Figure 5.10 Map of the hospital considered for performing propagation simulations, and the corresponding attenuations due to different walls [12].

5.5.2 Planning Simulator for "Smart Hospital"

A simulator has been developed in a Matlab environment, able to acquire a planimetry from an input .dxf file and perform propagation simulations [12]. The map area is decomposed into pixels, having a dimension equal to the simulation resolution, set to 0.1 m by 0.1 m. The simulation output is a database which is a collection of the power attenuations calculated for each of these pixels and for a given position of the base-station. As a reference, we have considered the planimetry described in the previous section. In the present analysis the Multi-Wall path loss model has been applied, and different types of walls are considered. Figures 5.11, 5.12, 5.13, and 5.14 show the path loss values evaluated by (5.7) extracted by running simulations while placing a single base-station in different locations within different types of rooms (patient, X-ray, MRT, and Operating rooms) on the hospital floor.

Results are presented both for the case of a patient room having plasterboard walls and the case of a MRT room with shielded walls. Therefore, reported figures are representatives of the coverage of each base station (fixed ZigBee transmitter) in the selected environment, for the considered transmission band, and provide us with an useful indication of the maximum coverage range achievable by a ZigBee transmitter operating in the band of interest. We observe that, a single base-station is able to cover more or less 10 patient-rooms according to its particular location, while for a shielded room (i.e. MRT room) the covered area remains confined only within the room.

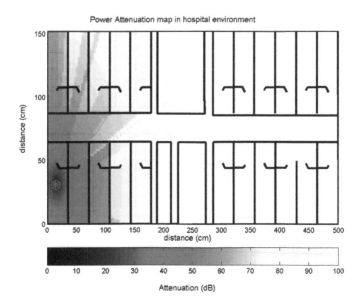

Figure 5.11 Path loss considering one fixed node placed within a patient room.

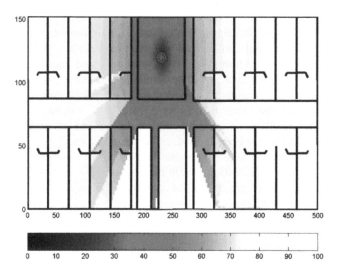

Figure 5.12 Path loss considering one fixed node placed within the X-ray room.

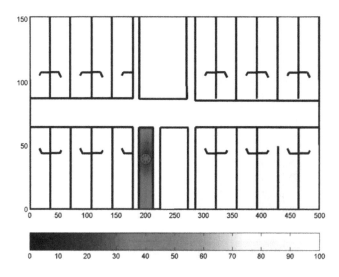

Figure 5.13 Path loss considering one fixed node placed within the MRT room.

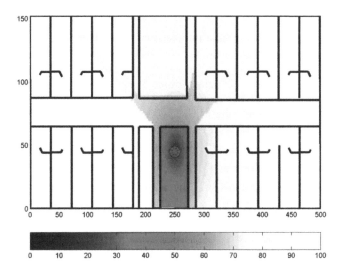

Figure 5.14 Path loss considering a fixed node within the Operating room.

5.5.3   Evaluation of Infrastructure Requirements for Positioning System

The minimum number of fixed nodes necessary to localize a target is evaluated as a function of the maximum Base Station Range r. In fact, as explained in the previous sections, localization of a target is effectively performed by a group of four base-stations. According to equations (5.6), (5.7), (5.8) and taking into account the propagation losses l_i introduced by a single type of wall i, the corresponding coverage radius is:

$$r = 10^{\left(\frac{P_{Tx} + G_{Rx} - S - \overline{PL_0} - l_i}{10\,n}\right)}$$

(5.9)

where G_{Rx} is assumed equal to 0 and $n = 2$, while P_{Tx} and S represent the transmitted power of the ZC and the receiver sensibility respectively. These values depend on the particular technology adopted. To quantify how many fixed devices are necessary to correctly exploit localization in the full hospital plan we analyzed snapshot simulations obtained for different positions of base-stations.

It emerges that shielded rooms, such as the MRT room or the operating room, need four nodes to be placed inside to perform localization, because the corresponding coverage radii are very small. On the contrary, the coverage radius for patient rooms is large enough that a node can cover a significant portion of adjacent room, while X-ray room is able to cover partially its closest patient-rooms.

In addition, we have to take into account the influence of the exact position of each node in each room to cover the maximum localization area as high as possible. To better understand this issue, we can consider, for example, a square area with side equal to the coverage range r, and we can firstly place the Base Station at the corner of the considered area as shown in the Figure 5.15. By Figure we can observe that the localized area, i.e. the area "seen" by the 4 Base Stations contemporaneously is a restricted center area, smaller than the total area to be covered r^2. It is obviously due to the limited coverage area of each single node which we can evaluate at the first approximation equal to $A_c < r^2$, and thus by considering the equation (5.5), the number of node needed for localization becomes:

$$N = 4\left\lceil \frac{A_T}{A_c} \right\rceil = 4\left\lceil \frac{A_T}{r^2} \right\rceil \tag{5.10}$$

To overcome we can consider the total area that we want to localize as a combination of sub-parts of square area with side of $r/\sqrt{2}$ as shown in the Figure 5.15. In this case each square area is completely localized by the Base Station, because each node is able to coverage the whole area, and thus we can assume the coverage area equal to the physical area $A_C = r^2/2$. This situation leads to a deep overlapping among adjacent area. It means that the equation (5.5) we can rewrite as:

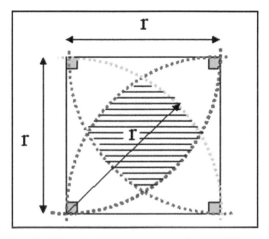

Figure 5.15 Localized area by 4 Base Station in a square area of r side [13].

$$N = 4 \left\lceil \frac{A_T}{A_c} \right\rceil = 4 \left\lceil \frac{A_T}{\frac{r^2}{2}} \right\rceil = 8 \left\lceil \frac{A_T}{r^2} \right\rceil \qquad (5.11)$$

By comparison of the equations (5.10) and (5.11) we can observe that in the latter situation there is a over-sizing of the nodes needed that can lead to a degradation of total performance due to both the increase of interference of adjacent nodes, and useless increase of the cost (see Figure 5.16).

A suitable combination of these two opposite cases can be to consider the total area to be localized as a square area of side of r, with Base Stations placed in the middle of each side instead of the corresponding corner (see Figure 5.17). In this new constellation the coverage area increases without change the dimension of the physical area to be monitored. For this case the number of nodes needed are evaluated according to the (5.10) equation with a consequence increase of the total performance of the localization system.

Therefore by taking into account the above consideration and the outcome of the simulations in each room of our case of study (see Figures 5.8, 5.9 and 5.10), we can evaluate the total number of nodes needed to perform localization in the analyzed area of Figure 5.7. Specifically, Table 5.2 shows the distribution of the Base Stations in each room, and then the total number to "localize" each tag in the floor of the hospital.

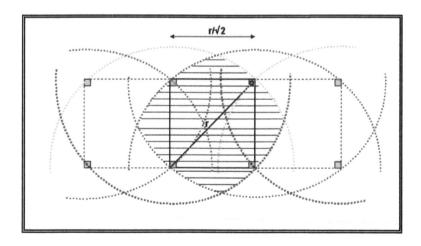

Figure 5.16 Localized area by 4 Base Station in a square area of $r / \sqrt{2}$ side [13].

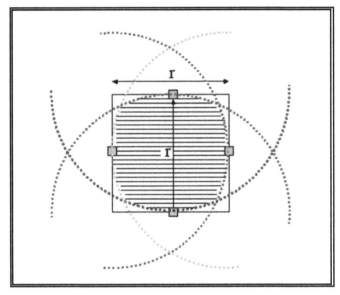

Figure 5.17 Localized area by 4 Base Station placed at the middle of each side, in a square area of side [13].

Table 5.2

Estimation of the number of the Base-Station needed for the Localization.

Parameter	Rooms	Fixed nodes
Patient	22	8
MRT	1	4
Operating	1	4
X-Ray	1	2
Total floor	1	18

5.6 CONCLUSIONS

The effects of indoor propagation of a ZigBee sensor network have been investigated. Specifically extensive RSSI measurements were carried out. The experiments were performed in a real environment, i.e. a not noise-free laboratory, in order to analyze the RF propagation behavior of ZigBee modules in real working condition, where they would normally operate when deployed. In addition a comparison of the experiment results with a propagation model has been also done. This study has been performed to verify the performance requirements of a WSN placed in real context to estimate what are the main factors that influence its efficiency for healthcare applications. Our analysis has led to consider not only external factors such as harsh environment (due to multipath, fading etc.), but even internal factors such as e.g. the hardware of the device.

We estimated the propagation taking into account a model for a hospital area, due to its restrictive characteristics. Some simulations and calculations have been carried out by considering the multi-wall propagation model that inlcudes the attenuation due to walls.

In the future new propagation models will be studied, and the influence of some factors, like the receiver sensitivity, the central frequency, or the presence of the people moving around, will be deeply analyzed.

References

[1]   I.F. Akyildiz et all. "A Survey on Sensor Networks" IEEE Communication Magazine, August 2002

[2]   ZigBee: "Wireless Control That Simply Works" page [Online]. Available: http://www.zigbee.org/en/resources/presentation.asp

[3]   E. Jafer et all. "A study of the RF characteristics for Wireless Sensor Deployment in Building Environment"IEEE SENSORCOMM 2009

[4]   M. Barrelet et all. "Effects of Antenna Polarization on RSSI Based Location Identification," ICACT 2009

[5]   J. Hightower et all. "Design and Calibration of the SpotON Ad-Hoc Location Sensing System," Seattle, WA, August 2001.

[6]   ZigBee: "ZigBee Wireless Sensor Applications for Health,Wellness and Fitness", white Paper, March 2009, http://www.zigbee.org/Markets/ZigBeeHealthCare/Overview.aspx

[7]   IEEE 802.15.TG4a page [Online]. Available: http://www.ieee802. org/15/pub/TG4a.html

[8]   T. M. Schfer et all. "Experimental Characterization of Radio Wave Propagation in Hospitals", Electromagnetic Compatibility, IEEE Transactions on Vol. 47, Issue 2, May 2005 Page(s):304 - 311.

[9]   Cushcraft Corporation "Antenna Polarization Considerations in Wireless Communications Systems", 1999-2002.

[10] T. S. Rappaport and T. Rappaport "Wireless Communications: Principles and Practice", 2nd Edition. Prentice Hall, Dec. 2001.

[11] "COST action 231, Digital mobile radio towards future generation systems final report", European Commission Brussels 1999.

[12] Integrated Project PULSERS PhaseII (No 506897) "D2a3-3 Definition of advanced UWB integration platforms - System Requirements – Final Version," January 2005

[13] UBISENSE, "Effects of psd limits on uwb positioning system practicability," Feb. 2005.

[14] Cushcraft Corporation "Antenna Polarization Considerations in Wireless

Communications Systems", 1999-2002.

[15] M. Barrelet et all. "Effects of Antenna Polarization on RSSI Based Location Identification," ICACT 2009

[16] M. R. Pellegrini, S. Persia, D. Volponi, G. Marcone, "RF Propagation Analysis for ZigBee Sensor Network using RSSI Measurements," WVITAE2011 - "2nd International Conference on Wireless Communications, Vehicular Technology, Information Theory and Aerospace & Electronic Systems Technology", Chennai, India, Febbraio 2011.

Chapter 6.

Dimensionality Reduction in Surface Electromyographic Signals for Pattern Recognition

Sujin Suwanna[1], Jirapong Manit[2] and Prakarnkiat Youngkong[2] (Ph.D.)

[1]Department of Physics, Faculty of Science, Mahidol University, Phayathai, Bangkok, Thailand
[2]Institute of Field Robotics, King Mongkut's University of Technology Thonburi, Bangkok, Thailand
sujin.suw@mahidol.ac.th; 52432302@study.fibo.kmutt.ac.th; youngkong@fibo.kmutt.ac.th

6.1 INTRODUCTION

This chapter provides an application of the dimensionality reduction technique to process surface electromyographic signals which can be used to determine gait-phase patterns. Dimensionality reduction is mainly aimed to represent high-dimensional data samples in a low-dimensional space so that most of relevant information in the data does not change. Usually, dimensionality reduction is processed before the signals can be used to extract other information or physical quantities. Preserving structures among the data points is a demand that must be met by ideal dimensionality reduction techniques, and this often becomes a measure which separates one technique from the others. Processing electromyographic signals, which are signatures of human gait cycles, can benefit greatly from this technique. For the purpose of pattern recognition such as those in gait-phase patterns, dimensionality reduction allows us to analyze, classify and visualize electromyographic signals accurately, which leads us to make appropriate correspondence between the signal patterns and gait cycles. In addition, with a large set of data, the dimensionality reduction technique can

Bio-Informatic Systems, Processing and Applications, 107-124

significantly reduce the computational time.

The main purpose of this chapter is to introduce some dimensionality reduction techniques commonly used in electromyographic signals processing, as well as related concepts in pattern extraction and recognition. An example of the application of a specific dimensionality reduction technique known as Neighborhood Component Analysis (NCA) is also presented at the end of the chapter, after which interested readers can find many useful references.

Surface electromyography (sEMG) is a method to record myoelectric signals which are formed by physiological variations in the states of muscle fiber membranes. In medical sciences, especially for rehabilitation purposes, it is well known that sEMG signals provide insights into how our muscles contract and relax. For instance, sEMG from specific muscle bundles can be used to characterize human gait cycles. In the field of engineering, the sEMG signals are widely used in many studies such as intelligent prosthetic and exoskeleton control inputs because of their ability in direct inflecting user's movement intention. Interested readers can find more details in References [1]–[4]. Improving the performance of sEMG pattern recognition is essential for theoretical study and for practical applications; especially those that require real-time feedback or control. Many schemes and techniques have been introduced to increase the pattern recognition accuracy and to reduce required computational time. This is indeed a very active research field; see References [4]–[10] for examples of research in this direction.

In this chapter, we present some techniques of dimensionality reduction, which have been generally used to improve pattern recognition and classification algorithms. This dimensionality reduction method is employed and implemented when we want to make a transformation from a high-dimensional space to a lower-dimensional space, while significant features of the original signals are still preserved.

After applying a dimensionality reduction technique, we can apply other signal processing techniques to a reduced set of data. For instance, a feature projection procedure can be used to separate and classify each class of patterns. It is hypothesized and thus far proven affirmative that a classification algorithm can work more efficiently. We will present precise definitions of separability and classification in a later section. It is common and useful to view the pattern recognition problem as consisting of three steps, namely, feature extraction, dimensionality reduction and classification [8]. For conceptual context, class separability and classification measures indicate how effective the chosen dimensionality reduction method is for the given task. The most suitable dimensionality reduction method is expected to yield the highest values of class separability as well as of classification accuracy.

In practice, sEMG signals must be collected from our muscles. For this, wireless sEMG signal recording equipment is used, for example, ZeroWire from Noraxon. In a typical sEMG measurement, electrodes for collecting sEMG signals are placed on many muscle bundles on both legs and other body parts.

Commonly, Ag-AgCl electrodes are used together with a 16-bit data acquisition at the sampling rate of 1500 Hz. Also, the band-pass filter between 10 Hz to 1000 Hz is applied. When a subject walks at a comfortable pace, the raw signals are collected and later analyzed. Such data acquisitions are generally studied in combination with the use of force plates to mark the duration of interested events. For instance, the stance phase interval will start when a signal from a force plate appears and will stop when value of the force signal is zero.

6.2 EMG-BASED PROSTHESIS CONTROL SCHEME

For a prosthesis-control scheme, sEMG signal has been used as system input data, because of its ability to directly inflect the user intentions. Three major components of the prosthesis control system for this scheme, including user, microprocessor and prosthesis, are shown in Figure 6.1. Wire or wireless sEMG is attached to the user body to collect signals from the target muscle groups. These signals will be collected continuously, and then sent to the microprocessor in the prosthesis to process through the block as shown in Figure 6.2. After signal pattern classification has been made, the microprocessor will use the result to control the actuator of the prosthesis. In this case, the knee prosthesis will need to be controlled by the damping of the knee joint while the user is walking on different gait phases. The over-all computational time should be less than 128 ms to make a good responsibility [11].

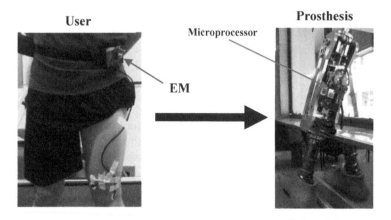

Figure 6.1 Components of sEMG-based prosthesis control system.

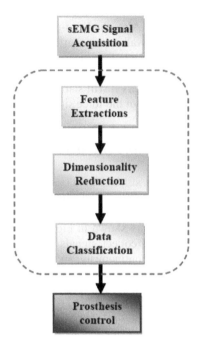

Figure 6.2 Block diagram of the sEMG signal pattern classification.

6.3 EMG FEATURE EXTRACTION

Since using raw sEMG signals is very difficult to classify movements or activities of muscle bundles, various types of feature extractions and their combinations have been introduced to classify sEMG signal patterns. These techniques principally extract only key features and are reliable in making feature correspondence with the muscle activities. Indeed, many techniques have been investigated by many researchers as can be found in References [5, 8, 15–17]. Similar to usual signal processing, the sEMG feature extraction can be done in both the time and frequency domains, usually with the Fourier transform as an agent to switch from one to the other. In this chapter, we will consider a few feature extraction methods in each domain. Only key equations are included here. For deeper details, readers are encouraged to look into, for examples, References [6, 8, 16–18].

6.4 TIME DOMAIN FEATURES

6.4.1   Mean Absolute Value (MAV)

MAV is one of the most popular time-domain feature extraction methods for sEMG signals. It is, in words, the average of the absolute values of the sEMG signals, where the average is taken over sampled values in one segment of the sEMG signals. In this case, MAV provides the amplitude of muscle contracting. As the name suggests, MAV is the average value of the magnitude of the signal. For definiteness, MAV is defined as

$$MAV = \frac{1}{N}\sum_{i=1}^{N}|x_i| \tag{6.1}$$

where N is the length of the signal segment, and x_i is the signal value of the ith sample. From equation (6.1), it can be seen that this averaging does not allow the offset of data values due to sign differences. In an ideal and unbiased sEMG signal collection, the arithmetic mean value of the signals is zero, as we believe that the muscles contract and relax symmetrically. In practice, the sEMG averaged value is normally found to be negligibly small. Consequently, MAV can be thought of as a measure of the averaged deviation of signals from zero, the ideal value.

6.4.2   Waveform Length (WL)

Waveform Length (WL) is defined as the cumulative length of the waveform over the time segment. It is related to the waveform amplitude, frequency and time duration via the equation:

$$WL = \frac{1}{N}\sum_{i=1}^{N-1}|x_{i+1} - x_i| \tag{6.2}$$

Thus, WL is a parameter which contains information on the waveform complexity in each segment.

6.4.3   Variance (VAR)

In a general statistical sense, the variance is the mean value of the square of the deviation, which is defined by

$$VAR = \frac{1}{N-1}\sum_{i=1}^{N}x_i^2 \tag{6.3}$$

Here, the average value of sEMG signals is taken to be zero, as in the ideal case. We notice that VAR uses the power of the sEMG signal as a feature. Often, we calculate the root mean square (RMS) value, defined in equation (6.4):

$$RMS = \sqrt{\frac{1}{N}\sum_{i=1}^{N} x_i^2} \qquad (6.4)$$

Approximately, the RMS value is the same as the standard deviation value of signals, which measures how much dispersion the data set inherits. The RMS value plays a central role in determining data cohesiveness, at least in a statistical sense. Moreover, for a large number of data points, the signal variation can be modeled as amplitude modulated Gaussian random process whose variance is given by the RMS. In electromyography, it is related to the constant force and non-fatiguing contraction.

6.4.4　Willison's Amplitude (WAMP)

Willison's Amplitude (WAMP) is defined as a number of times that the difference between sEMG signal amplitudes among two adjacent segments exceeds a pre-determined threshold. It is related to the firing of motor unit action potentials (MUAP) and the muscle contraction level. WAMP can be expressed as

$$WAMP = \sum_{i=1}^{N-1} f(|x_i - x_{i+1}|)$$
$$f(x) = \begin{cases} 1, & \text{if } x \geq x_{threshold} \\ 0, & \text{otherwise} \end{cases} \qquad (6.5)$$

Generally, the threshold value $x_{threshold}$ is chosen in the range of 10 mV and 100 mV, depending on the setting of the gain value of instrument [17]. In this study, the threshold value is set at 30 mV.

6.5 FREQUENCY DOMAIN FEATURES

Evidences from collected sEMG data have shown changes in the power spectrum of sEMG signals during muscle voluntary contraction. This is believed to be caused by muscle fatigue, which leads to a decrease in oscillation amplitudes. We are interested in investigating the changes in frequency to learn the activities of the muscle units, and thus perform various statistical analyses on sEMG signals in the frequency domain. Most notably, two commonly used spectral indicators are the mean frequency (MNF, also known as centroid frequency) and the median frequency (MDF), which are analogous to indicators in the time domain.

Statistically speaking, MNF is the first moment of the frequency, provided its distribution is known. As such, higher moments of frequency such as second moment (variance) or even fractional moments are possible as spectral indicators. In principle, indicators in the frequency domain compliment those in the time domain, and more information is gained by considering both together.

6.5.1   Median Frequency (MDF)

Median Frequency (MDF) is the median frequency, which is defined as the frequency such that the number of frequencies below it is equal to that above it. Thus, MDF satisfies the equation:

$$\sum_{j=1}^{MDF} P_j = \sum_{j=MDF}^{M} P_j = \frac{1}{2}\sum_{j=1}^{M} P_j \qquad . \qquad (6.6)$$

It should be noted that in the equation (6.6), the frequency is in discrete form, which may not be in case in general, i.e. the summation becomes an integration, but the meaning of MDF remains the same. In this case, the factor of 1/2 refers to the median, but we can also use other fractions such as 1/4 and 3/4 (interquartile frequencies).

6.5.2   Mean Frequency (MNF)

Mean Frequency (MNF) is the average value of the frequency, thus it is obtained from

$$MNF = \frac{\sum_{j=1}^{M} f_j P_j}{\sum_{j=1}^{M} P_j} \qquad (6.7)$$

where, in general, the summation can be integration and the values of P_j can be replaced by the distribution function of frequency.

In the case of voluntary contractions, frequency domain indicators are random variables with a particular probability distribution that generally depends on parameters such as the nature of the indicator, the signal length, the amount of time overlapping, and types of windows. The behaviors of MDF and MNF provide qualitative information of how the spectrum of the signals changes as a function of time. When they coincide, it suggests that the spectrum is symmetric with respect to the center. Otherwise, their difference indicates the skewness of the spectral distribution. In particular, if their ratio remains constant as a function of time, it suggests that the spectral scales are formed without changing the shape of the distribution.

6.6 DIMENSIONALITY REDUCTION TECHNIQUES

As an emphasis, a dimensionality reduction technique is an algorithm to reduce the number of features in the feature set, so that it facilitates the classification algorithm in the next step. In the dimensionality reduction step, several techniques have been introduced to handle sEMG signals.

6.6.1   Neighborhood Components Analysis (NCA)

This algorithm is a novel method proposed by J. Goldberger *et al.* [13] for learning a Mahalanobis distance measure used in the *k*-nearest neighborhood (KNN) classification algorithm. The Mahalanobis distance matrices can be represented by symmetric positive semi-definite matrices and estimated using inverse square roots. Instead of estimating the actual leave-one-out classification error of KNN, a more effective measure by using a differentiable cost function based on a stochastic neighbor assignment in the transformed space was introduced. Define the probability that a point *i* selects another point *j* as its neighbor (p_{ij}) as

$$p_{ij} = \frac{\exp(-\| Ax_i - Ax_j \|^2)}{\sum\limits_{k \neq i}^{M} \exp(-\| Ax_i - Ax_k \|^2)} \qquad , p_{ii} = 0 , \qquad (6.8)$$

where A is a transformation matrix. The objective of NCA is to maximize the expected number of points correctly classified under the scheme

$$f(A) = \sum_{i} \sum_{j \notin C_i} p_{ij} = \sum_{i} p_i \qquad . \qquad (6.9)$$

Here, the mathematical notation $\|\mathbf{x}\|$ means the vector norm or magnitude of the vector $\mathbf{x}$, and C_i is the set of points in the same class as i , i.e. $C_i = \{j: c_i = c_j\}$ where c_i is a class label.

6.6.2   Principal Component Analysis (PCA)

Principal Component Analysis (PCA) is an unsupervised linear dimensionality reduction algorithm using covariance matrix of the input data. This method seeks to embed data points in the lower dimensional space by constructing a subspace spanned by the orthogonal linear combination of basis vectors corresponding to the maximum-variance directions. This subspace is called the principal component (PC). The principal component which has the highest variance will be denoted as PC_1 and so on, descending to PC_n.

Since the variance depends on the magnitude of the input data, each data point must be offset by its mean value, thus shifted to a data set with zero mean. Suppose the original input data matrix is denoted by $\mathbf{X'}$, then the new input matrix $\mathbf{X}$ can be expressed as

$$\mathbf{X} = \mathbf{X'} - \mathbf{m} . \tag{6.10}$$

Then the covariance Σ of $\mathbf{X}$ is obtained from

$$\Sigma = \frac{1}{n} \mathbf{X} \cdot \mathbf{X}^T . \tag{6.11}$$

Using the eigen-basis decomposition, the covariance matrix Σ can be expressed as

$$\Sigma = \mathbf{U} \cdot \Lambda \cdot \mathbf{U}^T , \tag{6.12}$$

where $\Lambda = diag(\lambda_1, ..., \lambda_p)$ is the diagonal matrix of eigenvalues in ascending order. Here, $\mathbf{U}$ is the $p \times p$ orthogonal matrix containing eigenvectors. Accordingly, PC_{th} is obtained from the p_{th} row of the matrix $\mathbf{S}$, where

$$\mathbf{S} = \mathbf{U}^T \cdot \mathbf{X} . \tag{6.13}$$

Here, $\mathbf{U}^T$ acts as the weight matrix $\mathbf{W}$ which is used to project input data from the original space into a smaller subspace. That is,

$$\mathbf{S}_{k \times n} = \mathbf{W}_{k \times p} \cdot \mathbf{X}_{p \times n} \tag{6.14}$$

where $k < p$.

6.6.3   Linear Discriminant Analysis (LDA)

Linear Discriminant Analysis (LDA) is a linear supervised dimensionality reduction method, which seeks to maximize the linear discriminant of the data points in each class. It searches a set of vector, i.e. a linear mapping $\mathbf{M}$, whose linear combination yields the largest mean differences between classes, thus gains more linear discriminant. In this technique, two important matrices are constructed, namely, the within-class scatter matrix (S_W) and between-class scatter matrix (S_B). Mathematically, they are defined as

$$S_W = \sum_{j=1}^{c}\sum_{i=1}^{N_j}(x_i^j - \mu_j)(x_i^j - \mu_j)^T \text{ and}$$ (6.15)

$$S_B = \sum_{j=1}^{c}(\mu_j - \mu)(\mu_j - \mu)^T,$$ (6.16)

where c is the number of classes and N_j is the number of data points in the jth class. Here, x_i^j and μ_j are respectively the ith data value and the mean value in class j^{th}, and μ is the overall population mean. It should be noted here that, indeed, S_B is the covariance matrix of data in class, after subtracting the mean value vector in the class. Similarly, S_W is the summation of covariance matrices over all classes.

LDA will maximize the ratio between within-class scatter and between-class scatter by finding a linear mapping (M) such that

$$\Phi(M) = \frac{M^T \cdot S_B \cdot M}{M^T \cdot S_W \cdot M}$$ (6.17)

is maximized. If S_W is a nonsingular matrix, then $\Phi(M)$ is maximized when the column vectors of the projection matrix M are eigenvectors of $S_W^{-1}S_B$. The resultant matrix M therefore will project raw data X from the high dimensional space into a lower dimensional space, and satisfies the equation

$$Y_{k\times n} = M_{k\times p} \cdot X_{p\times n}$$ (6.18)

Both PCA and LDA have been applied successfully to study pattern recognitions in many situations. For instance, PCA has been used in face recognition, handprint recognition and robotics, so has LDA. Readers interested in these topics can find many useful references in Reference [12]. It is understandable that PCA is more suitable than LDA in some situations, while LDA is more in others. However, NCA is more modern, comparing with two classical algorithms PCA and LDA. For sEMG pattern recognition, we hypothesize that the NCA algorithm gives higher class separability than the other two, and thus initiate our investigation.

6.7 CLASSES SEPARABILITY MEASURE

A class separability measure is an indicator which can be used to discriminate between objects from different classes. Ideally, in computation, the more separable the classes are, the less extensive, the less complex and the less burdened the tasks will become. Consequently, we investigate and want to construct suitable class separability measures which can be used as indicators in signal processing of our sEMG data. The main assumption here is that patterns from the same class are close to each other (low separability) and those in different classes are relatively farther away (high separability).

There exist many separability measures, and their hybrids. However, in this section, we will focus on two class separability measures which are well known and suitable for use with sEMG.

6.7.1   Thornton's Separability Index (SI)

This is a measure of the degree to which inputs associated with the same output cluster together. It is shown in [18] to be an effective measure of class separability. The output value of SI ranges between 0 and 1. In the case that each class is in a well-separated cluster, the output will be close to 1. And, the index will approach to 0 when the clusters move closer. The SI is defined as:

$$SI = \frac{1}{n} \sum_i [(f(x_i) + f(x'_i) + 1] \bmod 2 , \tag{6.19}$$

where f is a binary target function, x'_i is the nearest neighbor of x_i and n is the number of points.

6.7.2   Fisher's Index (J)

Fisher's index measures the separability of the data cluster which has been projected into a new space. Its value J can be calculated by the ratio between traces of two class scatter matrices, between-class scatter matrix and within-class scatter matrix, as shown in equation (6.20)

$$J = \frac{Trace(\mathbf{S}_B)}{Trace(\mathbf{S}_w)} \tag{6.20}$$

where $\mathbf{S}_B$ and $\mathbf{S}_W$ are defined as in equations (6.15) and (6.16). From equation (6.20), it follows that J depends on $\mathbf{S}_B$, $\mathbf{S}_W$ and the number of classes.

We note that when data are highly clustered, the value of $Trace(\mathbf{S}_W)$ decreases. Likewise, when data are separated, the value of $Trace(\mathbf{S}_B)$ increases. When comparing the values of J from data set A with that from data set B, if J_A is higher than J_B, then it indicates that the data in set A is more dispersed than those in set B.

6.8 EXAMPLE OF APPLICATION OF DIMENSIONAL REDUCTION

In our study, we use high-end wireless EMG signal recording equipment, ZeroWire from Noraxon. sEMG signals are collected from four muscles on both legs; namely rectus femoris, biceps femoris, medial gastrocnemius and tibialis anterior, as shown in Figure 6.3. Using Ag-AgCl electrodes, the 16-bit data acquisition is performed at the sampling rate of 1500 Hz with the band-pass filter set between 10 and 1000 Hz.

Five healthy male subjects aged between 20–26 were instructed to walk barefoot at a comfortable pace. The walk path was 10 meters long on a hard floor implanted with three force plates which were used to identify the interval of swing phase and stance phase. Each subject repeated his walk five times. After data collection, the raw sEMG signals were labeled manually accordingly to the force signal. Figure 6.4 illustrates sEMG signals collected from a gait cycle. The classes of stance phase and swing phase signals are marked as 1 and -1, respectively. The whole data of each class is divided into two equal sets, a training set and a testing set for the classification.

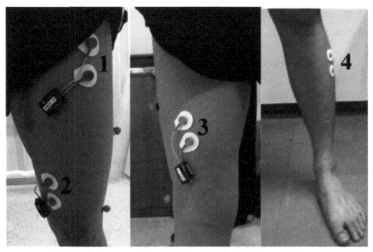

Figure 6.3 Electrodes location: (1) rectus femoris, (2) vastus medialis, (3) biceps femoris and (4) tibialis anterior.

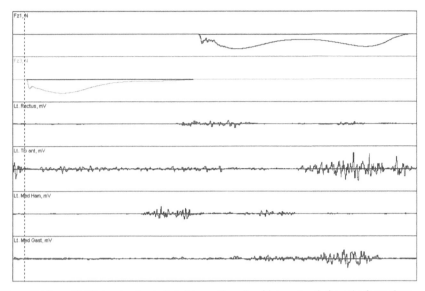

Figure 6.4 Raw sEMG signals of a single left leg walk cycle with two signals from the force plates. The top two signals on the graph are the force signals.

The features MAV, WL, VAR and RMS are calculated in a given segment of sEMG signals. According to the real-time control scheme presented in Reference [7], the performance of the analysis window length (T_a) between 32 ms to 256 ms does not yield significant differences. Thus, the T_a and window sliding time are selected to be 64 ms and 32 ms, respectively. In our calculation, the sample number N in a segment is set to equal 96.

After our dimensionality reduction is performed, we compare the values of class separability and classification accuracy. Our hypothesis in this study is that the class with high separability should also get higher classification accuracy. Our results are shown in Figure 6.5.

6.8.1   Class Separability

As shown in Figure 6.5, all extracted features are reduced from four-dimensional to two-dimensional space. The class separability can reveal the information of how easily these data sets can be separated. Considering just the SI in Figure 6.6, it can be seen that the separability of NCA in the MAV, RMS and WAMP is higher than those from other algorithms. Furthermore, the MAV and RMS features projected by NCA can gain the maximum SI score. This indicates that the stance phase and swing phase class of these two data sets are completely separated. So, MAV and RMS are good feature extractions for the NCA system. LDA and PCA get less average score than NCA. Regarding the variation of SI for each feature extraction, the dimensionality reduction algorithms are much effective to MAV, WL, VAR and RMS. The small variation of SI in WAMP can be explained that it is not much different for each algorithm to be applied with WAMP.

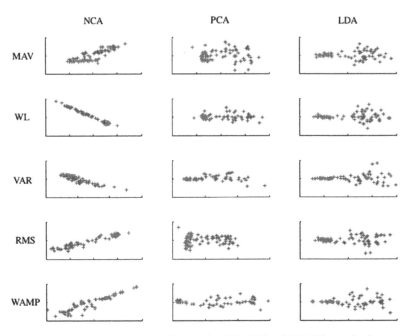

Figure 6.5 The two-dimension projected features by NCA, PCA and LDA. Blue marks denote stance phase while red marks denote swing phase.

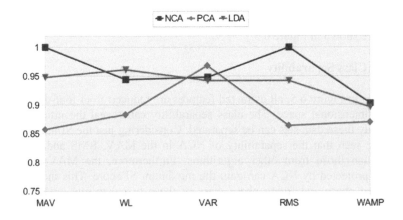

Figure 6.6 Thornton's separability index (SI) of each algorithm.

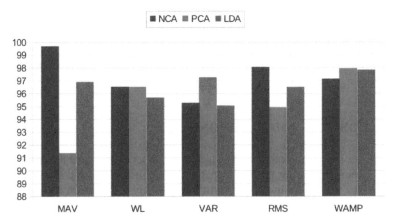

Figure 6.7 Classification accuracy of each projected features.

6.8.2   **Classification Accuracy**

As shown in Figure 6.7, each feature from the dimensionality reduction techniques can give an accuracy of more than 90 %. The most accurate system is the projected feature of MAV from NCA, with 99.7% accuracy results. And, the second most accurate system is also the RMS from NCA. From Table 6.1, it is seen that class separability of the sEMG data set relates to the classification accuracy.

6.9 CONCLUSIONS

We have investigated dimensionality reduction algorithms in combination with class separability index, and in particular, an application of NCA to reduce the dimension of the gait phase sEMG signals. After the feature extractions and dimensionality reduction steps, the pattern classification process improves its accuracy as expected. Our statistical results show that NCA yields better class separability. Furthermore, the average classification accuracy of the transformed

Table 6.1
Average Classification Accuracy and Separability Index
for each Dimensionality Reduction Algorithm

	NCA	**PCA**	**LDA**
Classification Accuracy	97.3%	95.6%	96.4%
Thornton's Separability Index	0.959	0.888	0.938
Fisher's Index	$6.06e^{-5}$	$2.15e^{-5}$	$2.58e^{-5}$

features by NCA is higher than those by PCA and LDA. Evidently, the NCA method can be used to perform dimensionality reduction in the data space for sEMG signals better than the other two techniques. Consequently, it yields more accurate classification and better pattern recognition between the stance phase and the swing phase with sEMG data.

6.10   PROBLEMS AND SOLUTIONS

This chapter investigates the dimensionality reduction techniques for applications in sEMG data processing to classify patterns of stance phase and swing phase in human gait cycles. Our question lies in the curiosity as to which dimensionality reduction method is best suitable for sEMG signals. The investigation began with two commonly used classical dimensionality reduction techniques, namely, PCA and LDA, but also considered the newer NCA technique. Our findings show that for sEMG signals NCA outperforms the other two considered techniques in both class separability measure and classification accuracy. This lays a foundation for our future research in pattern recognition in gait cycles with sEMG signals.

References

[1]   A. Delis, J. de Carvalho, G. Borges, S. de Siqueira Rodrigues, I. dos Santos, and A. da Rocha, "Fusion of Electromyographic Signals with Proprioceptive Sensor Data in Myoelectric Pattern Recognition for Control of Active Transfemoral Leg Prostheses," Engineering in Medicine and Biology Society (EMBC 2009 Annual International Conference of the IEEE), Sept. 2009, pp. 4755–4758.

[2]   J.-U. Chu, I. Moon, Y.-J. Lee, S.-K. Kim and M.-S. Mun, "A Supervised Feature-Projection-based Real-time EMG Pattern Recognition for Multifunction Myoelectric Hand Control," Mechatronics, IEEE/ASME Transactions on, vol. 12, no. 3, pp. 282–290, June 2007.

[3]   K. Momen, S. Krishnan, and T. Chau, "Real-time Classification of Forearm Electromyographic Signals Corresponding to User-selected Intentional Movements for Multifunction Prosthesis Control," Neural Systems and Rehabilitation Engineering, IEEE Transactions on, vol. 15, no. 4, pp. 535–542, Dec. 2007.

[4]   R. Khushaba, A. Al-Ani, and A. Al-Jumaily, "Orthogonal Fuzzy Neighborhood Discriminant Analysis for Multifunction Myoelectric Hand Control," Biomedical Engineering, IEEE Transactions on, vol. 57, no. 6, pp. 1410–1419, June 2010.

[5]   K. Englehart, B. Hudgin, and P. Parker, "A Wavelet-based Continuous Classification Scheme for Multifunction Myoelectric Control," Biomedical

Engineering, IEEE Transactions on, vol. 48, no. 3, pp. 302–311, March 2001.

[6]   J.-U. Chu, I. Moon, and M.-S. Mun, "A Real-time EMG Pattern Recognition Based on Linear-Nonlinear Feature Projection for Multifunction Myoelectric Hand," Rehabilitation Robotics, ICORR 2005 9th International Conference on, June 1–July 2005, pp. 295–298.

[7]   K. Englehart and B. Hudgins, "A Robust, Real-time Control Scheme for Multifunction Myoelectric Control," Biomedical Engineering, IEEE Transactions on, vol. 50, no. 7, pp. 848 –854, July 2006.

[8]   K. Englehart, B. Hudgins, P. A. Parker, and M. Stevenson, "Classification of the Myoelectric Signal using Time-frequency Based Representations," Medical Engineering & Physics, vol. 21, no. 6-7, pp. 431– 438, 1999.

[9]   M. Oskoei and H. Hu, "Support Vector Machine-based Classification Scheme for Myoelectric Control Applied to Upper Limb," Biomedical Engineering, IEEE Transactions on, vol. 55, no. 8, pp. 1956 –1965, Aug. 2008.

[10]   J. Manit and P. Youngkong, "Combination of Surface EMG Signal Features for Gait Phase Detection," Seventh Asian Conference on Computer-Aided Surgery, August 2011.

[11]   J.-U. Chu, I. Moon, S.-K. Kim, and M.-S. Mun, "Control of Multifunction Myoelectric Hand using a Real-time EMG Pattern Recognition," Intelligent Robots and Systems, IEEE/RSJ International Conference (IROS 2005), Aug. 2005, pp. 3511–3516.

[12]   A. Martinez and A. Kak, "PCA versus LDA," Pattern Analysis and Machine Intelligence, IEEE Transactions on, vol. 23, no. 2, pp. 228–233, Feb. 2001.

[13]   J. Goldberger, S. Roweis, G. Hinton, and R. Salakhutdinov, "Neighborhood Components Analysis," Advances in Neural Information Processing Systems 17. MIT Press, 2004, pp. 513–520.

[14]   L. Mthembu and J. Greene, "A Comparison of Three Class Separability Measures," Fifteenth Annual Symposium of the Pattern Recognition Association of South Africa, November 2004, pp. 63–67.

[15]   H. Hermens, T. B. van, C. Baten, W. Rutten, and H. Boom, "The Median Frequency of the Surface EMG Power Spectrum in Relation to Motor Unit Firing and Action Potential Properties," Journal of Electromyography and Kinesiology, vol. 2, no. 1, pp. 15–26, 1992.

[16]   H. H. D. Tkach and T. A. Kuiken, "Study of Stability of Time Domain Features for Electromyographic Pattern Recognition," Journal of Neuroengineering and Rehabilitation, vol. 7, no. 21, May 2010.

[17]   M. Zardoshti-Kermani, B. Wheeler, K. Badie, and R. Hashemi, "EMG Feature Evaluation for Movement Control of Upper Extremity Prostheses," Rehabilitation Engineering, IEEE Transactions on, vol. 3, no. 4, pp. 324–333, Dec. 1995.

[18] J. Greene, "Feature Subset Selection using Thornton's Separability Index and its Applicability to a Number of Sparse Proximity-based Classifiers," Twelfth Annual Symposium of the South African Pattern Recognition Association, 2001.

[19] D. Farina, R. Merletti and C. Disselhorst-Klug, "Multi-channel Techniques for Information Extraction from the Surface EMG," Electromyography, Physiology, Engineering, and Noninvasive Applications, Edit by R. Merletti and P. Parker, Wiley John Wiley & Sons, Inc., 2004, New Jersey.

[20] J. Goldberger, S. Roweis, G. Hinton, and R. Salakhutdinov "Neighborhood," Advances in Neural Information Processing Systems, 17, pp. 516.520, 2005.

Chapter   7.

Assessing a potential electroencephalography based algorithm during a monotonous train driving task in train drivers

Budi T Jap[1], Peter Fischer[2] and Sara Lal[1] (PhD)

[1]*Department of Medical and Molecular Biosciences, University of Technology, Sydney, New South Wales, Australia*
[2] *Signal Network Technologies, Pty Ltd, Lane Cove, New South Wales, Australia*
budi.t.jap@uts.edu.au, p.fischer@telstra.com, sara.lal@uts.edu.au

7.1   INTRODUCTION

Railway accidents have different causes, including poor track condition, faulty machinery, and human errors [1]. However, human error accounts for nearly 75% of train accidents, and most of the errors occur because crew members are fatigued [2,3]. Train drivers and controllers (or dispatchers) are normally required to work on a 24-hour irregular and rotating shift [4]. Shiftwork may interrupt the normal sleep-wake circadian cycle, which can lead to fatigue and performance impairment [5-8]. Fatigue is common among the train drivers and train traffic control crew members, with over half of the train drivers and traffic controllers severely fatigued during night shift [9]. Nearly a quarter of train drivers fall asleep while driving or waiting at station or stop signal during the day [10]. Several train accidents that occurred in New South Wales (NSW), Australia, in the recent years are related to fatigue and stress, including the Waterfall train accident in January 2003 [11]. Other countries, such as America and Canada, also report fatigue as a cause of several train accidents [11].

The irregular train driver shift may be managed through the use of fatigue

Bio-Informatic Systems, Processing and Applications, 125-138,

management software, such as the Fatigue Audit Inter-Dyne (FAID) software, which calculates the approximate fatigue level of train drivers and manages the train drivers' rosters to minimise the fatigue level that drivers experience while on the Job [12]. However, the software cannot account for drivers' activities outside the work hours, and hence, the FAID software's prediction of the drivers' fatigue levels when working may not be accurate [13]. Several train mishaps that occurred in Australia have been associated with fatigue, such as the Beresfield [14], Epping [15], Waterfall [16], and Benalla [13] train accidents. After such train accidents, FAID score was used in the post-accident investigation to identify pre-accident fatigue levels in the train drivers' and train guards'. The drivers involved in the Epping and Benalla accidents had FAID scores well below 80, which was the acceptable FAID score [13, 15]. However, fatigue was still believed to be the cause of the accidents [13, 15]. There are also reports of fatigue-related train accidents from other countries, such as USA [17, 18], and United Kingdom [19, 20].

The cost of train accidents not only affect those who are directly involved in the accidents, such as the drivers and the passengers, but also the community and the general population [21]. The direct costs of railway accidents mainly comprise lost of earnings for the victims and family, the pain and suffering experienced by the victims, family members, and the surrounding communities, and the property damage that the accidents may have caused [22]. Some of the costs of the accidents fall directly on the general public, such as loss of productivity due to workplace disruption when a person is unable to work after an accident. Other costs may affect the public indirectly in the form of an increase of freight costs due to accidents, or increase of costs of goods purchased by the general public [21]. Long-term psychological problems comprise some part of the cost of train accidents. After experiencing a train accident, approximately 16% of train drivers develop post-traumatic stress disorder, which can last for years [23-25]. Physical injury, such as restricted mobility in different parts of the body, disturbance of sight or hearing, and discomfort from scar or cartilage formations, also causes long-term problems for some of the victims [26]. The consequences of an accident to the victims can be long lasting and may affect them for a lifetime. Thus, there is an emotional burden on society as well as the financial burden.

In order to avoid fatigue related train accidents, all parties involved in the railway industries, such as drivers, controllers, and company managers, need to cooperate and have the same perception and strategy towards fatigue prevention [27, 28]. Education can allow better understanding that everyone could be prone to driving fatigue, and need to take rests regularly on a long journey [27]. Those who manage the working shifts need to schedule shifts that allow a longer resting period between shifts than the required amount of time to sleep to allow drivers to attend personal and social needs [11]. Although it is impossible to avoid night and early morning shifts, or irregularity in shift in a 24-hour railway industry, it is possible to adjust the shift lengths, the time between shifts and shift combinations in order to minimize fatigue [9]. However, it is not always possible to allow for such flexibility into the shift as the driver is compelled to work beyond the mental and physical capacity.

Several fatigue countermeasure techniques and devices have been researched and studied, and these may be useful to combat fatigue related train accidents. Automatic Train Protection (ATP) was developed to prevent trains passing signals at dangerous speeds, or failing to stop on terminating lines [29]. Others have developed fatigue countermeasure devices that detect eye closure, eye gaze, and blink rate measurement [30-32]. Correlation between saccadic eye movement velocity and pupil diameter with impaired driving performance exists [32]. However, studies have found that fatigue can occur in absence of eye closure and any visible physical signs of fatigue. Eye closure and blink rate may be too late to warn the driver of performance degradation due to fatigue, because the person may already have multiple micro-sleep periods before the physical signs of fatigue occur, such as eye blinking or closure [33, 34]. Some facial feature detection techniques have also been developed to observe the changes in facial expressions, such as yawning, glazed look, and frequent nodding, which can be used to detect fatigue [35, 36]. However, fatigue may occur with or without obvious appearances of drowsy behaviours, such as yawning, eye-blink, and nodding, and micro-sleeps may have occurred long before any physical appearance of drowsy mannerisms [34].

Electroencephalography (EEG) can be used to detect fatigue. Artaud et al. [37] found that EEG is one of the most reliable indicators of fatigue. The alert brain state has a mixture of alpha and beta activities [38]. An increase in alpha and theta activities accompanied by a decrease in beta activity can be interpreted as an indication of drowsiness [39, 40]. During the early stage of fatigue, the alpha activity may attenuate for a few seconds, and may reappear a few seconds later, and this alternating action may exist for a few minutes before alpha disappears at the onset of sleep [38, 39, 41]. An increase of theta and the change in alpha intensity has been associated with a deterioration of driving performance [42]. Jap et al. [43] have also shown significant decreases in beta activity as one gets fatigued during a monotonous train driving experiment.

Several studies have indicated that the use of combined EEG frequency bands may be a more reliable fatigue indicator [44-46]. The equation $(\theta+\alpha)/\beta$ can be used as a reliable fatigue indicator, since it combines the theta and alpha activities to detect changes in the alertness level [44]. The equation provides a ratio between the slow wave (θ and α) activities and the fast wave (β) activity. As one gets fatigued, the output of the equation is expected to show a significant increase in activity [43].

Hence, the aim of this study was to examine the changes in the equation $(\theta+\alpha)/\beta$ (identified as promising by Jap et al. [43, 47]) on 10 train drivers during a monotonous train driving task.

7.2 METHODS

Fifty male train drivers, aged 21-65 years (mean: 44 ± 9.4 years), were recruited to participate in the monotonous train driving study. All participants provided informed consent prior to entering the study, and held a current Rail Safety Worker Certificate (Driver) from the Department of Transport, Australia.

The study was conducted in a temperature-controlled laboratory. All participants were asked to refrain from consuming coffee, tea, or food approximately 4 hours prior to the study to minimise brain activity variation due to the effects food and beverage consumption [48]. Smoking was also restricted approximately 4 hours prior to the study, and participants were asked to refrain from alcohol consumption approximately 24 hours prior. Lifestyle Appraisal Questionnaire was used to establish selection criteria, that is, participants had to have no medical contraindication that could limit compliance, such as severe concomitant disease, alcoholism, drug abuse, and psychological or intellectual problems [49]. Approval for this study has been obtained from the institute's Human Research Ethics Committee (UTS HREC REF NO. 2006-176A).

7.2.1   Selection of the 10 participants

The Fatigue Likert scale, which was aimed to identify the immediate fatigue level, and was administered prior to and after the driving study, was utilised to identify participants that could be included in this case study. The range of score for the Fatigue Likert scale was from 1 (alert) to 4 (extremely fatigued).

Participants that were included in the current case study scored a 1 (alert) or 2 (slightly fatigued) prior to the monotonous driving session, and scored a 3 (moderately fatigued) or 4 (extremely fatigued) after the monotonous driving session. These participants would provide an observable brain activity transition from the alert state to the fatigued state.

7.2.2   Processing of EEG recordings

The EEG recording during the 30-minute monotonous train driving session was sectioned into 30 one-minute epochs. These epochs were then subjected to Fast Fourier Transform (FFT) to derive the four frequency components (delta (δ) (0–4 Hz), theta (θ) (4–8 Hz), alpha (α) (8–13 Hz), and beta (β) (13–35 Hz)) [50]. Figure 7.1 illustrates the FFT transformation process of the EEG recording into 1-second epoch.

After transforming the EEG recording into frequency domain, the EEG spectra were divided into 1-minute sections, and all epochs in the 1-minute section were averaged to obtain a value for each 1-minute section. This resulted in 30 one-minute average values for the 30-minute of monotonous train driving session. Figure 7.2 illustrates the segmentation of the driving session.

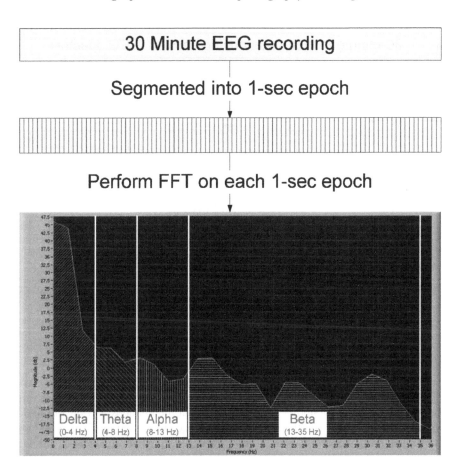

Figure 7.1 Transformation of time-domain EEG recording to frequency domain into one second epoch segments.

The first 5 minutes of the monotonous driving session was calculated as the alert baseline and used as a comparison baseline against the EEG activity for the remaining driving session.

7.2.3   Selection of Algorithm

Previous studies by Jap et al. [45, 51] and Eoh et al. [44] have shown that a combination of EEG frequency bands could be used to detect the transition from the alert state to the fatigue state. A combination of theta, alpha, and beta frequency bands in the equation $((\theta+\alpha)/\beta)$ has been shown to increase as the driver starts getting fatigued. This equation provides a ratio between slow wave EEG activities (theta and alpha) and fast wave EEG activity (beta). Slow wave EEG activities has been shown to increase during fatigue [41, 44], while fast wave activity decreases during fatigue [52, 53].

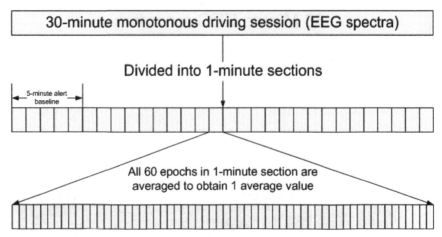

Figure 7.2 Segmentation of monotonous driving session into one-minute averages.

Therefore, the algorithm that will be used for this simulation study is the equation $((\theta+\alpha)/\beta)$.

7.3 RESULTS

Ten train driver volunteers, aged 36-47 years (mean: 42 ± 7.6 years), were included in this case study. The total average driving time for the train drivers were 32 ± 2.9 minutes. Studies have indicated that monotonous driving for 20-30 minutes could lower alertness level and induce driver fatigue [54, 55]. One volunteer (volunteer 7) was extremely fatigued during the monotonous driving session, and the study was stopped after 24 minutes of monotonous driving. Table 7.1 presents the age, driving duration, and the pre- and post-study Fatigue Likert score of the 10 train drivers that were included in this case study.

From Table 7.1, the volunteers can be separated into three groups based on the pre- and post-study Likert scale. The first group is for those volunteers (volunteers 1, 4, 5, and 10) who were in an alert state (Likert scale 1) at the beginning of the study, and were in a moderately fatigued state (Likert scale 3) at the end of the study. The second group consists of volunteers (volunteers 2, 6, and 9) who were alert (Likert scale 1) prior to the study, and were extremely fatigued (Likert scale 4) after the study. The last group consists of volunteers (volunteer 3, 7, and 8) who were slightly fatigued (Likert scale 2) at the beginning of the study and were extremely fatigued (Likert scale 4) after the study.

The combination of theta (θ), alpha (α), and beta (β) activities in the equation $(\theta+\alpha)/\beta$ was expected to increase as the individual gets fatigued [44]. An increase of theta activity and a decrease of beta activity have been associated with deterioration of driving performance due to fatigue [42-44, 46, 56].

Table 7.1
Age, Driving Time, and Pre- and Post-Study Fatigue Likert Score

#	Age	Total Driving Time (minutes)	Pre-study Likert	Post-study Likert
1	47	32	1	3
2	42	32	1	4
3	44	32	2	4
4	43	31	1	3
5	40	32	1	3
6	36	32	1	4
7	38	24	2	4
8	46	31	2	4
9	38	34	1	4
10	41	35	1	3

7.3.1   Volunteer group 1

The first group of volunteers consists of those who reported to be alert (Likert scale 1) at the beginning of the driving session and were moderately fatigued (Likert scale 3) at the end of the driving session (volunteers 1, 4, 5, and 10). The result of the $(\theta+\alpha)/\beta$ activities for this group showed a 100% increase from the 5-minute baseline activity (Figure 7.3). Volunteers 1 and 10 had a steady increase

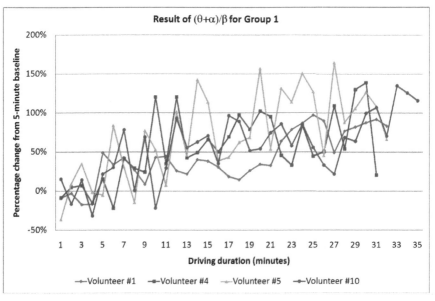

Figure 7.3 Volunteers 1, 4, 5, and 10: frontal activity; showing percentage change from a 5-minute baseline average; θ = theta; α = alpha; β = beta.

of activity throughout the driving session, while volunteers 4 and 5 had large fluctuations of $(\theta+\alpha)/\beta$ activity throughout the driving session, which averaged to approximately a 100% increase from the 5-minute baseline activity at the end of the driving session.

7.3.2   Volunteer group 2

The group of volunteers, who were alert (Likert scale 1) at the start of the driving session and were extremely fatigued (Likert scale 4) at the end of the driving session, showed a higher increase of $(\theta+\alpha)/\beta$ activity as a result. At the end of the driving session, volunteers 2 and 6 showed that $(\theta+\alpha)/\beta$ activity was 130%-160% higher than the 5-minute baseline (Figure 7.4). Volunteer 9 showed a greater increase of 200% in $(\theta+\alpha)/\beta$ activity from the baseline activity.

7.3.3   Volunteer group 3

The third group of volunteers consisted of those who were slightly fatigued (Likert scale 2) at the beginning of the driving session and extremely fatigued (Likert scale 4) after the driving session. This group of volunteers showed a similar trend as the first group that reported to be alert (Likert scale 1) prior to the driving session and was moderately fatigued (Likert scale 3) at the end of the

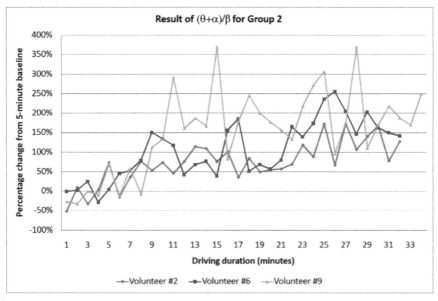

Figure 7.4 Volunteers 2, 6, and 9: frontal activity; showing percentage change from a 5-minute baseline average; θ = theta; α = alpha; β = beta.

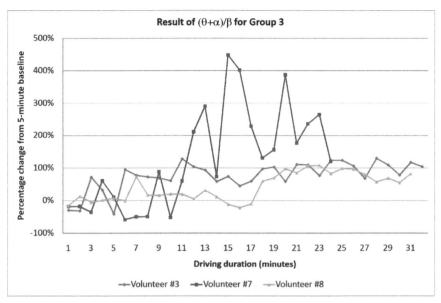

Figure 7.5 Volunteers 3, 7, and 8: frontal activity; showing percentage change from a 5-minute baseline average; θ = theta; α = alpha; β = beta.

driving session. Volunteers 3 and 8 had a steady increase of activity which was approximately 100% higher than baseline (Figure 7.5). However, volunteer 7, who was extremely tired throughout the driving session, had a greater increase of approximately 200% higher than the baseline activity after 15 minutes of driving monotonously.

7.4 DISCUSSION

Train driving can be considered a complex task, since it relies heavily on several aspects of cognitive tasks, such as recognition, decision-making, object detection, and sustained attention [57-59]. Drivers always have to be aware of the surrounding environment in order to act promptly and accurately in an emergency. However, when a driver is fatigued, the capacity to drive safely will be reduced significantly [59]. Fatigue has a strong influence to reduce awareness of the surrounding area, because it affects the physical and mental capacity to perform work [NTC, 11, 60]. Fatigue during driving is associated with longer response times, increased difficulty in identifying and processing important information from the surrounding environment, lapses in attention, and more frequent errors [58].

The most effective countermeasure against drowsiness and performance degradation is to have enough sleep or rest before driving [61-63], but this may not always be possible. When an individual feels fatigued while driving, the best practical action to prevent fatigue related accident is to stop driving and take a break for approximately 30-minutes, which is helpful to regain awareness and alertness [34, 64]. A 15-minute nap, followed by caffeine intake, during the 30-minute break is extremely helpful to reduce sleep propensity [34, 64]. However, it may not always be practical and possible to take a nap break while driving a train.

As long as drivers acknowledge that their ability to drive has decreased and take a rest when feeling drowsy, fatigue related accidents can be avoided [65]. Drivers should also abstain from monotonous driving during their normal sleeping periods or when they feel uncertain of being able to drive safely. However, most drivers will ignore the signs of fatigue, and hence, these practical countermeasures have major shortcomings. Direct technological fatigue countermeasures will be extremely helpful to alert drivers of early onset of fatigue.

Although automatic fatigue countermeasure devices will assist in detecting fatigue and preventing fatigue related accidents, these devices will need to be highly reliable and accurate in detecting and alerting fatigue [66]. When fatigue countermeasures are in operation, some drivers may feel secure with these devices installed in their devices, and may ignore obvious signs of fatigue as long as the devices 'remain silent' [66].

Electroencephalography (EEG) can be used to detect fatigue and has been found to be quite reliable [37]. A good test and retest reliability and high reproducibility of EEG activity has also been found during two episodes of monotonous driving sessions for the delta and theta bands [67], while others have reported acceptable reliability for alpha and beta activities [68, 69].

Beta activity recorded from an EEG during a monotonous driving task has been shown to decrease significantly when an individual is fatigued [43, 44]. According to Belyavin & Wright [52], the most sensitive indicator of fatigue is the significant decrease in beta activity, although others have argued that an increase in alpha activity is the most reliable indicator of reduced vigilance [70]. A decrease in beta activity has been associated to worsening performance and reduced vigilance [41, 44, 52].

The current study have investigated the percentage change in activity for the output of the equation $(\theta+\alpha)/\beta$ proposed by Brookhuis & Waard [46] and Eoh et al. [44]. Equation $(\theta+\alpha)/\beta$ combines both theta and alpha powers to detect changes in the alertness level [44]. Our studies and those of others have found that theta activity increases as one becomes fatigued [41, 44, 45]. The activity resulted from the output of equation $(\theta+\alpha)/\beta$ has been shown to increase as the individual gets fatigued, as a result of a decrease in beta activity and an increase in theta [44, 51]. Eoh et al. [44] believed that this equation, which combined alpha and theta powers, could be a basis for a more reliable indicator of fatigue, especially "during the repetitive phase transition between wakefulness and micro sleep" [44].

The result of the current study for the equation $(\theta+\alpha)/\beta$ shows an increasing trend as the individuals get fatigued. At the end of the driving session, the resultant activity for the equation increased by approximately 100% when compared to a 5-minute baseline average for the individuals who reported to moderately fatigued after the driving session. This increase was greater (between 150%-200%) for those who reported to be extremely fatigued at the end of the driving session. For the volunteers who were slightly fatigued at the start of the driving session, the percentage increase of the results of equation $(\theta+\alpha)/\beta$ was approximately 100%, since the baseline average activity might already be higher due to a pre-existing slightly fatigued state. The result of the current study is similar to the findings that other studies have reported in the literature [44, 45, 51].

This study has shown that monitoring EEG activity changes would be useful as a fatigue countermeasure device. A 100% increase of the activity output from the equation $(\theta+\alpha)/\beta$ would be a sign that the individual is moderately fatigued. A combined effort of these two EEG activity changes would confirm whether or not the individual is moderately fatigued. However, when the individual has commenced driving while in a fatigued state, the EEG baseline may not always reflect the correct EEG alert activities. This may produce an incorrect result, since the individual may already be extremely fatigued when the 100% increase of output from equation $(\theta+\alpha)/\beta$ activity occurred subsequently.

Some studies proposed a fatigue detection technique through the use of neural network or wavelet transform to classify alertness and drowsiness [38, 41, 71]. While others proposed techniques using Independent Component Analysis (ICA) algorithm to remove EEG artefacts as well as estimating drowsiness [7, 72]. These techniques may be combined with detecting the change in EEG activities to improve the detection rate. As Heitmann et al. [73] has observed that no single technology or technique is reliable enough for detecting driver fatigue, there is the need to combine different technologies or techniques together in order to increase reliability and reduce false alarms. The drawbacks of one technique may be compensated with the strength of the others.

7.5 CONCLUSION

The current study has investigated the changes of the resultant activity as an output of equation $(\theta+\alpha)/\beta$, proposed by Brookhuis & Waard [46] and Eoh et al. [44], on ten train drivers. There are three different groups of drivers that were included in this case study. The first group consisted of those who were alert at the start of the driving session, and were moderately fatigued at the end of the driving session. This group of drivers showed about 100% increase in activity from equation $(\theta+\alpha)/\beta$ at the end of the driving session. The second group reported to be alert prior to the driving session, and were extremely fatigued at the end of the driving session. The result of this group was approximately 150%-200% increase in activity derived from equation $(\theta+\alpha)/\beta$ at the end of the session. The last group of drivers began the monotonous driving session in a slightly fatigued state, and completed the driving session in an extremely fatigued state.

An increase of approximately 100% in activity from equation $(\theta+\alpha)/\beta$ was recorded for this last group.

Future studies may need to look at combining the above algorithm $((\theta+\alpha)/\beta)$ with other algorithms, such as the use of neural network, wavelet transform, or ICA algorithm, to detect changes in EEG activities and to increase reliability and reduce false alarms, as suggested by Heitmann et al. [73].

References

[1] V. Esslinger, R. Kieselbach, R. Koller et al., "The railway accident of Eschede - technical background," Engineering Failure Analysis, vol. 11, no. 4, pp. 515-535, 2004.

[2] G. D. Edkins, and C. M. Pollock, "The influence of sustained attention on Railway accidents," Accident Analysis & Prevention, vol. 29, no. 4, pp. 533-539, 1997.

[3] G. J. S. Wilde, and J. F. Stinson, "The monitoring of vigilance in locomotive engineers," Accident Analysis & Prevention, vol. 15, no. 2, pp. 87-93, 1983.

[4] J. Dorrian, S. D. Baulk, and D. Dawson, "Work hours, workload, sleep and fatigue in Australian Rail Industry employees," Applied Ergonomics, vol. 42, no. 2, pp. 202-209, 2011.

[5] D. F. Dinges, "An overview of sleepiness and accidents," Journal of Sleep Research, vol. 4, no. Supplement 2, pp. 4-14, 1995.

[6] P. Thiffault, and J. Bergeron, "Monotony of road environment and driver fatigue: a simulator study," Accident Analysis & Prevention, vol. 35, no. 3, pp. 381-391, 2003.

[7] C.-T. Lin, R.-C. Wu, S.-F. Liang et al., "EEG-based drowsiness estimation for safety driving using independent component analysis," IEEE Transactions on Circuits and Systems I: Regular Papers, vol. 52, no. 12, pp. 2726- 2738, 2005.

[8] S. A. Ferguson, N. Lamond, K. Kandelaars et al., "The Impact of Short, Irregular Sleep Opportunities at Sea on the Alertness of Marine Pilots Working Extended Hours," Chronobiology International: The Journal of Biological & Medical Rhythm Research, vol. 25, no. 2/3, pp. 399, 2008.

[9] M. Härmä, M. Sallinen, R. Ranta et al., "The effect of an irregular shift system on sleepiness at work in train drivers and railway traffic controllers," Journal of Sleep Research, vol. 11, no. 2, pp. 141-151, 2002.

[10] A. Austin, and P. D. Drummond, "Work problems associated with suburban train driving," Applied Ergonomics, vol. 17, no. 2, pp. 111-116, 1986.

[11] National Transport Commission (NTC), Fatigue Management Within The Rail Industry: Review of Regulatory Approach, National Transport Commission (NTC), Melbourne, Australia, 2004.

[12] A. Fletcher, and D. Dawson, "A predictive model of work-related fatigue based on hours of work," Journal of Occupational Health and Safety -

Australia and New Zealand, vol. 13, no. 5, pp. 471-485, 1997.

[13] Australian Transport Safety Bureau (ATSB), Derailment of Train 5MB7 at Benalla VIC, NSW Department of Transport, Transport Safety Bureau, Canberra, 2007.

[14] Australian Transport Safety Bureau (ATSB), Coal Train Collision Beresfield NSW 23 October 1997, NSW Department of Transport, Transport Safety Bureau, Canberra, Australia, 1998.

[15] Australian Transport Safety Bureau (ATSB), Collision Between Suburban Electric Passenger Train 1648 and Suburban Electric Empty Train 1025, NSW Department of Transport, Transport Safety Bureau, Canberra, 2003.

[16] P. A. McInerney, Special Commission of Inquiry into the Waterfall Rail Accident, vol. 1, Governor of the State of New South Wales, Sydney, NSW, Australia, 2005.

[17] National Transport Safety Board (NTSB), Railroad Accident Brief: Side Collision of Burlington Northern Santa Fe Railway Train and Union Pacific Railroad Train near Kelso, Washington, November 15, 2003, DCA-04-MR-003, National Transport Safety Board (NTSB), Washington DC, USA, 2005.

[18] National Transport Safety Board (NTSB), Railroad Accident Report: Collision of Union Pacific Railroad Train MHOTU-23 With BNSF Railway Company Train MEAP-TUL-126-D With Subsequent Derailment and Hazardous Materials Release, Macdona, Texas, June 28, 2004, NTSB/RAR-06/03, National Transport Safety Board (NTSB), Washington DC, USA, 2006.

[19] Rail Accident Investigation Branch (RAIB), Collision at Badminton - 31 October 2006, Department of Transport, UK, 2007.

[20] Rail Accident Investigation Branch (RAIB), Dispatch of a train with unsecured load, Basford Hall Yard, Crewe - 21 February 2006, Department of Transport, UK, 2007.

[21] Bureau of Transport and Regional Economics (BTRE), Rail accident costs in Australia, 108, Department of Transport and Regional Services, Commonwealth of Australia, Canberra, 2002.

[22] Bureau of Transport and Regional Economics (BTRE), Costs of Rail Accidents in Australia - 1993, Department of Transport and Regional Services (DOTARS), Canberra, ACT, Australia, 1995.

[23] R. Farmer, T. Tranah, I. O'Donnell et al., "Railway suicide: the psychological effects on drivers," Psychological Medicine, vol. 22, no. 2, pp. 407-414, 1992.

[24] A. Jabez, "Back in control," Nursing Times, vol. 89, no. 39, pp. 46-47, 1993.

[25] S. Karlehagen, U. F. Malt, H. Hoff et al., "The effect of major railway accidents on the psychological health of train drivers--II. A longitudinal study of the one-year outcome after the accident," Journal of Psychosomatic Research, vol. 37, no. 8, pp. 807-817, 1993.

[26] A.-L. Andersson, L.-O. Dahlbäck, and P. Allebeck, "Psychosocial consequences of traffic accidents: a two year follow-up," Scandinavian

Journal of Social Medicine, vol. 22, no. 4, p

Chapter 8.

Detecting Driver Drowsiness with Examples using EEG and Body Movement

Stacey Pritchett (PhD)[1,2], Eugene Zilberg (PhD)[2], Zheng Ming Xu (PhD)[2], Murad Karrar (PhD), David Burton (PhD)[2] and Sara Lal (PhD)[1]

[1]*Department of Medical and Molecular Biosciences, University of Technology Sydney, Broadway, NSW, Australia*
[2]*Compumedics Pty Ltd, Abbotsford, Victoria, Australia*
ezilberg@compumedics.com.au

This chapter provides a general overview of the characteristics of drowsiness, and the current methods used in the detection of drowsiness in drivers. This is followed by a more in depth discussion on two examples of drowsiness detection through the use of spontaneous electroencephalography (EEG). The methods of EEG analysis presented are that of spectral analysis and a pattern recognition style. The analysis styles chosen in these examples were designed to improve the sensitivity of detecting short-lived patterns in the EEG. The pattern recognition style also results in a convenient manner to further analyse the underlying properties of the existing patterns, beyond what is capable with spectral analysis. The addition of the body movement signals to the EEG analysis is assessed to investigate the increased strength of association between driver drowsiness and its physiological predictors through the use of more than one independent signal source.

8.1   INTRODUCTION

The terms drowsiness and fatigue are often used interchangeably in the literature, and refer to the state of lack of concentration due to sleepiness [1-3]. May and Baldwin [4] have indicated that fatigue can either be sleep or task related. Sleep related fatigue is influenced by the natural circadian rhythm, where periods of increased drowsiness occur from 2 – 6am and then again from 2 – 4pm. Sleep related fatigue is also influenced by quality and duration of sleep and wake-time after sleep. Task related fatigue is influenced by the amount of time spent on a task and how intensive or monotonous that task is.

Driving when drowsy or fatigued reduces driver performance to a similar extent as driving when intoxicated [3]. Therefore, it is not surprising that driver fatigue is attributed to 14-18% of fatal accidents in Australia [5], resulting in more than 60 road deaths and 300-400 serious road injuries per year in Victoria alone [3]. Furthermore, the risks of experiencing drowsiness when driving is increased with increased time on task [4], meaning that drowsiness and fatigue can be an increased work safety hazard for professional drivers (car, truck or train) and pilots. Therefore, the detection of drowsiness in drivers is an important aspect of future in-vehicle safety measures.

8.2 THE CHARACTERISTICS OF DRIVER DROWSINESS

There are different ways that drowsiness can manifest itself in a driver. These include the visual indications of the driver's movements or mannerisms, a reduction in the driving performance and the changes of the driver's physiological signals.

8.2.1   Visual charateristics

Wierwille and Ellsworth [6] verified the ability of a trained rater to visually score drowsiness level from the recorded video of subjects undertaking a driving task in a simulator. From this they were able to demonstrate a characteristic set of visual indications of drowsiness as described in the following paragraph.

As drowsiness increases, eye blinks and movements progress from being "normal" fast blinks and focus, to having increased rate and duration of blinks. There is increased loss of apparent focus, and a decreased degree of eye opening, otherwise known as "heavy eyelids". Facial tone slowly decreases, and body movements progress from the normal occasional head, arm or body movement onto restlessness in the very early stages of drowsiness. In later stages of drowsiness there are periods that are void of any body movements, followed by large or abrupt corrective movements. These periods of lack of movement increase in length and frequency with increasing severity of drowsiness.

These characteristics have been further investigated by Vural *et. al.* [7], who verified the characteristics as described by Wierwille and Ellsworth. They also discovered that while yawning increases with drowsiness, in the period of extreme drowsiness, about 1 minute prior to falling asleep, the rate of yawning actually reduces again.

8.2.2   Characteristics of driving performance

Fairclough and Graham [8] investigated the effects of sleep deprivation on driving performance. The investigation monitored lateral, longitudinal and speed control, in subjects that had a full nights sleep, partial sleep deprivation (4hrs sleep), full sleep deprivation (no sleep) and a full nights sleep, but over the legal blood alcohol limit.

Partial sleep deprivation can be assumed to be representative of moderate drowsiness and full sleep deprivation relates to severe drowsiness. The lateral control measures Fairclough and Graham used were a) lane crossing, which is when 2 of the vehicles wheels crossed either the outside lane boundary or the centre line; b) near crossing, which is when the lateral movement of the vehicle indicated that unless corrective measures were taken, a lane crossing would occur within 2 seconds; and c) steering wheel reversal rate, which is where zero crossings indicated the rate of corrective steering movements. The longitudinal control measure was the amount of "time" left between the vehicle being driven and the one in front of it. It was measured by dividing the distance between the two vehicles by the speed they were travelling. The speed control measure was the standard deviation of the vehicles speed. These measures were recorded every 5 minutes as either a rate of occurrence, or an average.

The results of the longitudinal control measures investigated by Fairclough and Graham, indicated that when drivers know they are drowsy, they tend to leave extra room between themselves and the vehicle in front of them to compensate. The variability in speed was greater with any length of sleep deprivation than with the influence of alcohol. The lateral control results indicated that any length of sleep deprivation reduced the rate of corrective steering wheel movements, where the influence of alcohol did not. With partial sleep deprivation, the rate of lane crossings was not significantly greater than the control group of a full nights sleep, however the rate of near crossings was significantly greater than any of the other groups. With full sleep deprivation, the rate of lane crossings was significantly greater than that of partial sleep deprivation or the control group, and was similar to that of the alcohol group.

Although Fairclough and Graham did not look at the variability of the vehicle position compared to the lane boundaries in time (lane position), their results imply that the variability of lane position would increase with drowsiness. Their results demonstrated that as drowsiness increases, the reduction in vehicle control, and therefore driver performance, progresses from low risk behaviour (i.e. increased speed variability and reduced corrective steering movements) with moderate drowsiness, to high risk behaviour (i.e. increased rate of unintentional lane departures) with severe drowsiness.

8.2.3   Characteristics of physiological signals

Physiological signals, such as respiration rate, the electrocardiogram (ECG) and electroencephalogram (EEG), are also known to vary with drowsiness.

The ECG is the measurement of the electrical activity of the heart. The ECG features 3 distinctive wave complexes (P, QRS and T) as illustrated in Figure 8.1. The P-wave is generated by the activity of the atriums of the heart. The QRS complex is generated as the ventricles are preparing to contract, and the T wave is generated during ventricular contraction (or ventricular systole) [9].

It was initially suggested that heart rate variability, measured as the mean square of heart rate in a 40s interval, was statistically associated with drowsiness onset, although unreliably [10]. Since then, the ECG has been further analysed, and a set of characteristics related to increasing drowsiness has been developed.

Work by Takahashi and Yokoyama [11] shows that as drowsiness increases, the heart rate decreases, and the beat to beat variability of heart rate increases. They have demonstrated the increase in heart rate variability through the use of measures such as the coefficient of variation of the R-R intervals (standard deviation/mean*100), and the changes in the duration of the ventricular contractions normalised with respect to the R-R intervals (R-T interval / √ R-R interval). These investigators also demonstrated that respiration rate decreases with high levels of drowsiness and the regularity of breathing pattern also decreases, as yawning and intermittent larger breaths occur.

The EEG is a recording of the electrical activity of the brain. It has long been used to monitor sleep [12], and is the main indicator of the occurrence of microsleep episodes [13]. As drowsiness is a pre-curser to sleep, the EEG has been suggested as a means to be able to directly measure fatigue and drowsiness as well [14].

In interpreting the EEG, there are 4 standard frequency bands; delta, theta, alpha and beta. Figure 8.2 shows examples of the 4 frequency bands, as recorded from a sleep study.

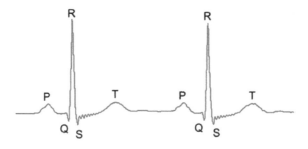

Figure 8.1 The structure of the ECG showing two consecutive heart beats, highlighting the P wave, QRS complex and T wave.

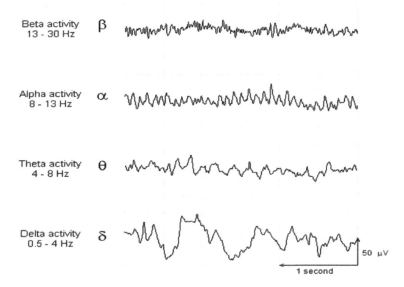

Figure 8.2 Standard EEG rhythms, as seen during a sleep study.

Each of these frequency bands is indicative of different neurological states, as summarised bellow [15]:

- The beta rhythm occurs at frequencies from 13 – 30 Hz and is present in the background EEG of most people during the awake state.
- The alpha rhythm occurs at frequencies from 8 – 13 Hz and is most frequently seen during the awake, relaxed state with eyes closed. Alpha waves are attenuated with periods of increased attention and during eye opening. However, the appearance of alpha activity when eyes are open is a sign of drowsiness, and is a phenomenon known as alpha burst [16].
- The theta rhythm occurs at frequencies from 4 – 8 Hz and is often indicative of drowsiness, representing the slowing of alpha rhythms.
- The delta rhythm occurs at frequencies from 0.5 – 4 Hz and is generally only clearly present during the deeper stages of sleep.

From this brief summary it can be seen that as drowsiness increases, the occurrence of alpha and theta activity in the EEG also increases. The generators relating to the alpha rhythm seen during the resting awake state with eyes closed are different to the alpha seen during drowsiness [17], and in a study by Torsvall and Akerstedt [18] it was noted that even in a subject that did not show clear alpha in the resting EEG, there was an increase in alpha activity with increasing subjective sleepiness.

8.3 MONITORING DRIVER DROWSINESS

The ultimate goal of drowsiness detection in the transport operator is to enable an early warning system that will alert the driver to the need to take counter measures before driving performance deteriorates to a safety critical level. This can be achieved by monitoring some, or all, of the typical indicators of driver drowsiness in order to estimate current drowsiness level. As there are different indicators of drowsiness and fatigue, there are different approaches that can be taken to estimate and detect the level of drowsiness or fatigue being experienced by the driver in question.

There are three main signal types used in the literature for the detection of driver drowsiness: 1) vehicle-based performance measures, 2) visual features and 3) physiological measurements. These relate to the typical characteristics of drowsiness in drivers as discussed earlier. Each approach to monitoring driver drowsiness has its own individual advantages and limitations associated with it [1]. A common conclusion when current methods of driver drowsiness detection are reviewed is that the use of a combination of the different indicators increases the reliability of the drowsiness detection [1, 19-21]. However, for the purpose of this review, the three main signal types will be addressed separately.

8.3.1 Monitoring using vehicle-based performance

As was discussed earlier, driving performance such as lane-tracking, steering wheel movement, and speed control are all effected characteristically by increased driver drowsiness. Therefore it is possible to estimate driver drowsiness level using these indicators without having direct monitoring of the driver in question, making this monitoring style particularly unobtrusive to the driver.

Sayed and Eskandarian [22] have reported 89.9% accuracy in identifying drowsiness versus non-drowsiness using angle data of the steering wheel. Variations in the steering wheel angle reflect the style of corrective steering movements associated with maintaining the vehicle's position with respect to the centre of the driving lane. In their work, the recorded angles of the steering wheel were pre-processed to remove angles relating to curves in the simulated road, thereby resulting in data that reflected only the corrective movements. The recorded angles were then discretised into eight-dimensional vectors that could be fed into an 8-layer input, 2-layer output artificial neural network for driver drowsiness estimation. This method of drowsiness detection appears to utilise relatively straight forward and easily implemented data acquisition, however the need to pre-process the data to correct for road curvature indicates that real-life application of this would require more sophisticated measures providing data relating to the layout of the road being travelled on. This could perhaps be provided by using image-tracking of the road surface to predict road curvature.

Image-tracking of the road surface would not only provide a means to accurately process the corrective movements of steering, it would also provide a direct method to monitor the lateral vehicle control measures relating to lane position, such as the standard deviation of lane position, unintentional lane departure, and time to lane departure. Fletcher *et al* [23] demonstrated the use of video image to detect lane boundaries and therefore vehicle control. They highlighted the importance of an adaptable selection of the processing method in image analysis, as the different processing methods each have ideal conditions for use. Their method combines visual cues and physical constraints to estimate the lane boundaries from the video image. Using a Monte Carlo sampling algorithm allows an optimised selection of the estimated lane boundaries and up-coming lane curvature based on Bayesian theory. Although Fletcher *et al* have included sample images of the lane detection indicating robustness to shadows, obstacles and multiple line markings on the road surface; they have not shown any statistical validation of their algorithm.

Another potential method of tracking road boundaries includes the use of radar. An example of this has been provided by Wijesoma, Kodagoda, and Balasuriya [24], who have utilised a 2 dimensional ladar (laser radar) measurement system. This system relies on the existence of curbs or gutters in order to detect the road edge, and is unable to provide information regarding lane position. Therefore this method of tracking road boundaries would be more suited to detecting obstacles on the road, but it would have the possibility of increasing the accuracy of tracking lane position when used in combination with image-processing techniques.

8.3.2   Monitoring using visual features

From reviewing the available literature on automated detection of driver drowsiness, it is apparent that the most widely researched monitoring method is the use of image processing of the visual features of a driver, specifically the eyes. Perhaps this is not surprising considering the 1996 Report from the National Highway Traffic Safety Administration [25] suggested that a measure of percentage of time the eyes are at least 80% closed in a given period (PERCLOS) was a highly rated real-time measure of alertness.

There are many approaches to the detection of eyelid movement from video images, and the following is just a small sample of these methods. In separate research by Liu *et al* [26], Danisman *et al* [27] and Wu et al [28], initially the region of video image that contained the face was defined using Haar-like features with cascaded classifiers (an image processing algorithm provided in the OpenCV library http://opencv.willowgarage.com/wiki/). Liu *et al* [26] then used the difference between successive images to identify eyelid movements, and the defining of eyes opened or closed was based on the change in the amount of grey in the pixels. Using controlled blinking patterns Liu *et al* demonstrated a 98% accuracy of detecting blinks. Danisman *et al* [27] have used a neural network-based eye detector to locate the eyes within the face area, and once the eyes were located and rotated to sit horizontally, a test of horizontal symmetry of the pixels in the eye region was used to determine eyes open or closed. They showed 94.8% accuracy of detecting eye blinks. Wu *et al* [28] achieved 85.2 – 95% accuracy of detecting blinks using a radial-symmetry transform with a support vector machine classifier to determine the location of the eye, and a local binary pattern method based on the level of grey in the pixels with a support vector machine classifier to determine eyes open or closed. Although these researchers did not test their algorithm on real-life drowsiness, it is clear that it is possible to monitor blinking pattern reliably using video images.

8.3.3   Monitoring using physiological measurements

Traditional physiological measurements, such as pulse waves [29, 30], ECG [29], or EEG [31, 32] can all be used to monitor driver drowsiness. However, it is also possible to use devices such as accelerometers [33], infra-red sensors [34] or piezoelectric transducers [35] to obtain less common measurements, such as body or head movements, which also show characteristic signs of drowsiness.

Lee *at al* [34] demonstrated the use of infra-red sensors in the head-rest of the driver's chair to monitor head movements. Based on feature extraction of indicators such as sudden movement (like would occur with a "head drop" episode) or gradual change in distance between head and head-rest, they have shown that the use of head movement data alone has a 61 – 78% accuracy of detecting drowsiness periods.

Takahashi and Yokoyama [11] investigated changes in parameters of the ECG and respiration during drowsiness in drivers. They tested 14 indices in a linear regression to find the relation between these parameters and the self-assessed level of drowsiness. From this, they found that Heart Rate, root Mean Square of Successive Differences, deceleration-related heart rate variability, R-T interval and normalised duration of the ventricular contractions were all statistically significant ECG parameters. The mean power frequency and respiration instability were the statistically significant respiration parameters of the multi-variate linear regression. Results indicated that these parameters and self-assessed drowsiness levels were highly correlated (correlation co-efficient of 0.719), but the relationship was weakly linear ($R^2 = 0.397$). Although the results published by Takahashi and Yokoyama are promising in relation to the ability to use ECG and respiration signals to monitor driver drowsiness, it may prove difficult to record these signals in real-life applications. In separate work by Ramesh, Nair and Kunnathu [30], and by Hu *et al* [29], the heart beat is being utilised via detection of pulse waves. Ramesh, Nair and Kunnathu [30] record the pulse wave via IR sensors embedded in the steering wheel (pulse oximetry recording). Although they have shown that the system they have designed is able to accurately record the pulse wave form, they have not elaborated on how they detect driver drowsiness using this signal beyond that they use heart rate variability. Hu *et al* [29] have similarly used the steering wheel to record the pulse wave, however they have captured the mechanical movement and changes in contact pressure caused by the pulse in the grip on the steering wheel. By using two layers of piezoelectric film with a layer of foam sandwiched between, the top layer of piezoelectric records both the grip and the small variations in the grip caused by the pulse, and the bottom layer records only the over-all grip. Therefore, pulse can be extracted from the signal via adaptive filtering. Hu *et al* have also described how they have utilised the heart rate variability from the pulse wave to monitor drowsiness. The heart rate time series is analysed in the frequency domain using autoregressive method and the power density spectrum is divided into the high frequency band (HF = 0.15-0.4 Hz) and low frequency band (LF = 0.04-0.15 Hz). The results that Hu *at al* presented show that the ratio between these two bands (LF/HF) decreases as drowsiness increases. They have not included statistical validation of the detection of drowsiness level though.

EEG alpha activity is reactive to both eye closure and drowsiness separately, and is reported to occur in bursts during drowsiness [16, 32, 36]. Picot *et al* [31] demonstrated the use of relative alpha power to detect drowsiness. They estimated the EEG power spectrum every 1s using the Short Time Fourier Transform (STFT) of 2s of EEG. From this they calculated the relative alpha and theta power for every 1s, prior to smoothing this data via a median filter. The median filter was calculated across 10 data points (equivalent to 10s, progressing 1s at a time) and enabled the spurious data caused by artifacts to be removed from the relative power data. Next a Means Comparison Test (MCT) on 30 data points (equivalent to 30s, progressing 1s at a time) was used to normalise the relative power data with respect to a baseline value that was calculated from the first 60s of the record. Results from Picot *et al* established that normalised relative alpha power alone was a stronger predictor of drowsiness than if data from the theta band was included in the analysis. They demonstrated an 85% true positive and 20% false positive detection rate of drowsiness. Although these are good results, the duration of alpha bursts are in the range of up to a few seconds, and analysis that relies on comparatively long time intervals (i.e. 30 – 60s), may not adequately reflect the more subtle changes occurring in the EEG with the lighter levels of drowsiness.

Karrar *et al* [37] indicated that the analysis of the morphological features of alpha bursts provided greater detection accuracy of transition to drowsiness than the use of conventional spectral analysis. More recently, Simon *et al* [32] have published work on the detection of alpha bursts (referred to as alpha spindles in their work) though the use of the spectral density plot. In the work by Simon *et al*, Fourier transforms were performed on 1s epochs of EEG (with a 750ms overlap), in a manner that increased the parameters of the individual alpha bursts that could be investigated. From the spectral density plot, a peak frequency was found between 3-40Hz, and if it lay within the alpha band, the spectral plot was further analysed to determine if this peak was due to an alpha burst. The parameters investigated were the spindle frequency, amplitude, duration and rate (all calculated from the amplitude spectral density graph). Using a driving simulation where drivers have self-assessed end of session due to experience of excessive drowsiness, Simon *et al* demonstrated that spindle rate (number of spindles or bursts in a given time period) showed a 90% increase between the first 20 minutes and the last 20 minutes of driving, compared to alpha power, which only showed a 32% increase. Their work indicates that the alpha burst parameters are more responsive to drowsiness than alpha power parameters, and correlates with the results shown by Karrar et al [37]. However in order to determine the presence of an alpha burst (or spindle), Simon *el al* [32] have defined a 'noise' level on the spectral density graph to compare the spectral peaks against, which was calculated retrospectively from data from the entire recording. Therefore their current methods, as published, are not suitable for real-time application.

8.4   EXAMPLES OF DETECTING DROWSINESS USING SPONTANEOUS EEG AND BODY MOVEMENT MEASUREMENTS

In this section, two examples of driver drowsiness detection using the spontaneous EEG are presented. The EEG has long been used to monitor sleep, and automatic sleep classification (staging) from the EEG is a common practice. A part of automatic sleep staging is the identification of sleep spindles, which are similar in morphology to the alpha burst patterns that are seen during drowsiness. Therefore, the following examples of alpha burst detection for the purpose of driver drowsiness detection   includes both EEG power analysis, and a pattern recognition analysis derived from current methods of detecting sleep spindles during automatic sleep staging.

It is known that the use of the power spectrum averaged across time diminishes the detection of sparse short-lived alpha burst patterns [36], therefore the power spectrum analysis presented in the following utilises the maximum relative alpha power rather than the average relative alpha power of a given segment in time. The use of pattern recognition analysis of the EEG in the time domain was selected as the second example of drowsiness detection, as this analysis style improves the sensitivity of detecting short-lived patterns in the EEG. It also results in a convenient manner to further analyse the underlying properties of the existing patterns, beyond what is capable with spectral analysis.

As there is always a level of uncertainty in utilising a single signal source for the detection of drowsiness, the following examples also demonstrate the increased association between physiological indicators and drowsiness level through the use of a secondary indicator: body movement.

Data used in these examples was collected from 60 (44 male) non-professional drivers holding a current licence, aged between 20-60 years. EEG from channels C_4 and O_2 was collected, along with body movement data from the driver's chair, as participants were driving a car simulator at the Monash

Table 8.1
Trained Observer Rating (TOR) scale, based on the Wierwille drowsiness continuum [6]

Drowsiness		Video Image Indicators
Level	State	
0	Alert	Normal fast eye blinks, often reasonably regular; Apparent focus on driving with occasional fast sideways glances; Normal facial tone; Occasional head, arm and body movements.
1	Slightly drowsy	Increase in duration of eye blinks; Possible increase in rate of eye blinks; Increase in duration and frequency of sideway glances; Appearance of "glazed-eye" look; Appearance of abrupt irregular movements – rubbing face/eyes, moving restlessly on the chair; Abnormally large body movements following drowsiness episodes; Occasional yawning.
2	Moderately drowsy	Occasional disruption of eye focus; Significant increase in eye blink duration; Disappearance of eye blink patterns observed during alert state; Reduction on degree of eye opening; Occasional disappearance of facial tone; Episodes without any body movements.
3	Significantly drowsy	Discernable episodes of almost complete eye closure, eyes never fully open; Significant disruption of eye focus; Periods without body movements (longer than for level 2) and facial tone followed by abrupt large body movements.
4	Extremely drowsy	Significant increase in duration of eye closure; Longer duration of episodes of no body movement followed by large isolated "correction" movements.

University Accident Research Centre (MUARC), Melbourne. For further details of the study methods, refer to [35, 38].

8.4.1   Selection of a gold standard of drowsiness

In order to validate the effectiveness of a chosen indicator for monitoring driver drowsiness, it is important to choose a suitable base-line or 'gold standard' of actual experienced drowsiness.

In the examples presented here, video of the subjects during simulated driving sessions was recorded to serve as a reference for drowsiness level. Drowsiness level was visually assessed using the Trained Observer Rating (TOR).

The TOR is based on the Wierwille drowsiness continuum [6] and has 5 levels of drowsiness: Alert, Slightly drowsy, Moderately drowsy, Significantly drowsy and Extremely drowsy (ranked 0-4 respectively, Table 8.1). Drowsiness level was assessed using consecutive 10 second intervals of video footage.

Figure 8.3 (a) shows the progression of drowsiness in an individual during a simulated driving session that lasted almost 1 hour, as assessed using the TOR scale. Figure 8.3 (b) shows the relative alpha power of this same individual. From these figures it appears that the relative alpha power increases with higher levels of drowsiness as assessed using the TOR. As it is known that alpha power increases with increasing drowsiness [31], this indicates the validity of the use of the TOR score as a suitable base-line for comparison.

In the following examples, when association between indicators and drowsiness level were assessed, the TOR score was used as an average over 30 seconds.

When the performance of the developed drowsiness detection algorithms was assessed, each driving session was classified into segments consisting of 3 different categories:

- *Reference segments* – the first 15 minutes of driving for each experiment session served as a reference period for calculating reference alpha power and burst values

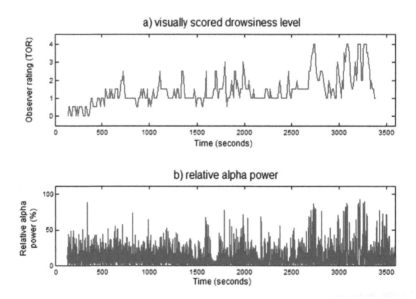

Figure 8.3 Progression of drowsiness during a simulated driving session, a) as assessed by the TOR scale, and b) the relative alpha power of the EEG.

- *Drowsy segments* - the duration from the transition-to-drowsiness point (TOR changes from < 3 to ≥ 3) until 60 seconds after the last test-point after the transition to maintain TOR ≥ 3
- *Alert segments* –all segments other than the reference or drowsy segments

8.4.2   Using the spontaneous EEG to detect drowsiness

In these examples, initially the association between the EEG indicators and driver drowsiness were established, and then the drowsiness detection algorithms were statistically validated. Then body movement data was added to the EEG indicators to test for strengthening of association between drowsiness level and its physiological indicators.

8.4.2.1   Spectral Analysis

Most often, spectral analysis of the EEG is achieved through the use of the Fast Fourier Transform (FFT). The equation of the FFT that is used in this example is as follows, where X_k is the transformed measurements in the frequency domain, x_n is the original measurements in the time domain, N is the total number of observations, n is the observation number ($n = 0$ ... $N-1$) and k is the frequency observed ($k = 0$ ... $N-1$) [39].

$$X_k = \sum_{n=0}^{N-1} x_n e^{-\frac{2\pi}{N}nk}$$
(8.1)

In this example, the raw EEG data was filtered to remove frequencies bellow 0.3Hz and above 35Hz. Then the Fourier Transform was calculated using the FFT function in Matlab, using a moving window of 1 second length with a 0.75 second overlap, giving a power spectral density array every 0.25 seconds.

Relative alpha power (rel_{alpha}) was calculated for each power spectral density array using the following equation, where Pow_{8-13Hz} is the power values from the spectral density array relating to the alpha frequency band ($8 – 13Hz$) and $Pow_{0.5-30Hz}$ is the power values from the spectral density array relating to the whole EEG frequency band ($0.5 – 30Hz$).

$$rel_{alpha} = \frac{\sum Pow_{8-13Hz}}{\sum Pow_{0.5-30Hz}}$$
(8.2)

A reference level of alpha power (ref_{alpha}) was calculated from the first 15 minutes of the driving session as the average relative alpha power for that 15 minute segment.

Table 8.2
Parameters of the linear regression model for average drowsiness predicted from the maximum relative EEG alpha power

Covariate	Regress. Coeff.	t-value	p-value	R^2
$Max_{adjusted}$ (C_4)	1.248 [1.037; 1.406]	11.57	<0.001	0.181
$Max_{adjusted}$ (O_2)	2.012 [1.83; 2.194]	21.66	<0.001	

For the remainder of the driving session, spurious data caused by artifacts was minimised in the relative alpha power data by averaging the relative alpha power over consecutive 2.25 second segments (6 rel_{alpha} data points). The maximum alpha power (Max_{alpha}) was then calculated for every 30 seconds as the maximum of the averaged rel_{alpha} data points in that 30 second period (20 averaged data points). An adjusted maximum alpha power ($Max_{adjusted}$) was then calculated for each 30 second segment of EEG as follows.

$$Max_{adjusted} = Max_{alpha} - ref_{alpha} \qquad (8.3)$$

The adjusted maximum alpha power was calculated for both the C_4 and O_2 channels of EEG, and a multivariate linear regression model was used to establish the association between the alpha power and driver drowsiness. The adjusted maximum alpha power from C_4 and O_2 were used as the independent variables to predict the average drowsiness level of a 30 second period, as assessed by the TOR scale. Table 8.2 provides a break-down of the regression coefficients, t and p-values of the linear regression. Although these results indicate a relatively poor linear relationship between the alpha power and drowsiness level ($R^2 = 0.181$), a statistically significant association was established ($p < 0.05$). Therefore, a drowsiness detection algorithm based on the use of the maximum alpha power was developed and tested.

The maximum alpha power was used to form a drowsiness detection algorithm as follows.

The alpha power drowsiness detection algorithm calculated the binary drowsiness index (0=alert and 1=drowsy) using consecutive intervals of 1 minute. A drowsiness threshold ($Threshold_{spectral}$) was established as shown in (8.4), where $K_{spectral}$ is the cut-off parameter of the drowsiness classification for spectral analysis, and has a direct impact on sensitivity/specificity. ref_{alpha} is the reference level of alpha power, which was calculated in the manner discussed earlier. Through analysis, $K_{spectral}$ was found to have a range of $0.03 - 0.2$, with a default value of 0.1 which provided the best sensitivity/specificity combination.

$$Threshold_{spectral} = \frac{K_{spectral}}{(1 + e^{-(2 \times (ref_{alpha} - 1.5))})} \qquad (8.4)$$

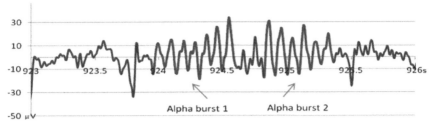

Figure 8.4 Example of alpha bursts on background EEG.

A weighted maximum alpha power (Max_{weight}) was calculated each minute by finding the average maximum alpha power from the two 30 second intervals within that minute (Max_{alpha1} = *the maximum alpha power from the first 30 seconds; and* Max_{alpha2} = *the maximum alpha power from the second 30 seconds*).

$$Max_{weight} = (Max_{alpha1} + Max_{alpha2}) / 2 \tag{8.5}$$

An adjusted weighted maximum alpha power ($Max_{weight-adj}$) was then calculated using the reference level of alpha power (ref_{alpha}) which was described earlier.

$$Max_{weight-adj} = Max_{weight} - ref_{alpha} \tag{8.6}$$

The adjusted weighted maximum alpha power of both C_4 and O_2 channels was then compared to the drowsiness threshold, and a value of 1 (drowsy) was assigned to any periods where either EEG channel produced a $Max_{weight-adj}$ above the threshold.

In order to test the sensitivity/specificity of the alpha power to detect drowsiness, TOR ≥ 3 was selected as the threshold for presence of drowsiness, as discussed in section 8.4.1. Each driving session was divided into drowsy and alert periods and statistical analysis was carried out in the same manner as is described in [38]. The AROC value was found to be 0.719 with a sensitivity/specificity of 0.72/0.95, indicating a good performance of the maximum alpha power in detecting drowsiness.

8.4.2.2   Pattern Recognition

Figure 8.4 shows an example of 2 individual alpha bursts in a 3 second sample of spontaneous EEG.

Individual parameters that were thought to describe the characteristic of the alpha burst were derived within each burst. The eight alpha burst parameters that were included in the analysis for this example were:

- *Burst duration* – the duration of the alpha burst

Table 8.3

Parameters of the linear regression model for average drowsiness predicted from the EEG alpha burst measures

Covariate	Regress. Coeff.	t-value	p-value	R^2
Burst duration (C_4)	0.043 [0.036; 0.05]	11.49	<0.001	
Burst duration (O_2)	0.067 [0.058; 0.075]	15.59	<0.001	
Mean amplitude (C_4)	-0.009 [-0.011; -0.007]	-7.77	<0.001	
Mean amplitude (O_2)	-0.008 [-0.011; -0.006]	-6.18	<0.001	
Wave duration variance (C_4)	0.792 [0.582; 1.003]	7.39	<0.001	0.271
Slope smoothness measurement (O_2)	1.908 [1.437; 2.378]	7.59	<0.001	
Relative amplitude (C_4)	0.457 [0.388; 0.527]	12.85	<0.001	
Relative amplitude (O_2)	0.413 [0.35; 0.476]	12.8	<0.001	

- *Current alpha wave count* – the number of contiguous alpha waves in the current alpha burst
- *Minimum alpha wave count* – the minimum number of contiguous alpha waves to be considered as an alpha burst
- *Mean amplitude* – the mean of the peak to peak amplitude of the individual waves in the alpha burst
- *Relative amplitude* – the ratio of the mean peak to peak amplitude during the alpha burst period compared to that of the 2 second period prior to the start of the alpha burst
- *Wave duration variance* – the standard deviation derived from all the durations between two consecutive maximum or minimum peaks in the alpha burst
- *Wave similarity* – a measure of the degree of similarity between the individual waves in the alpha burst, comparing factors such as individual wave amplitudes, duration, etc.
- *Slope smoothness measurement* – a measure of the contribution of background noise to the signal.

To establish the association between the alpha burst parameters and driver drowsiness, the above parameters, as recorded from the C_4 and O_2 channels, were used as the independent variables in a linear regression to predict the average drowsiness level of a 30 second period, as assessed by the TOR scale. Using 4 waves as the minimum alpha wave count and including only significant variables as covariates, the goodness of fit for this model was $R^2 = 0.271$. A break-down of the regression coefficients, t and p-values of the linear regression including only the significant variables are provided in Table 8.3.

Although these results indicate a relatively poor linear relationship between the alpha burst parameters and drowsiness level, this relationship has a stronger linearity to it than that of maximum alpha power. Again, a statistically significant association between these parameters and drowsiness level was established (p < 0.05), and it is likely that the poor linear relationship is due to the fact that alpha bursts can have two sources, early drowsiness and eye closure. It is possible therefore, that the relationship between alpha bursts and drowsiness is not linear, however for the purpose of detecting drowsiness onset, a linear relationship is not essential.

Figure 8.5 provides a flow diagram of the algorithm that was developed to detect drowsiness onset using the alpha burst parameters.

The *burst duration* was found to have the strongest association with drowsiness, as indicated by the t values (Table 8.3), and therefore the alpha burst drowsiness detection algorithm was designed with *Burst duration* as the definitive parameter. The other parameters that were utilised in the drowsiness detection algorithm were *Current alpha wave count, Minimum alpha wave count, Wave duration variance* and *Slope smoothness measurement*, which were employed to weight the outcome based on how robustly the alpha bursts resemble a pure alpha pattern.

The alpha burst drowsiness detection algorithm calculated the binary drowsiness index (0=alert and 1=drowsy) using intervals of a minimum of 1 minute, with 10 second overlap. Interval time was extended to include the entirety of any alpha burst that crossed over either the start or end of the 1 minute period.

Reference values for the alpha burst algorithm were initially calculated from the first 15 minutes of the driving session. A threshold value (*Threshold$_{burst}$*) for the detection of drowsiness was calculated from this reference section, according to (8.7). If no alpha was present during this time, then default values were utilised.

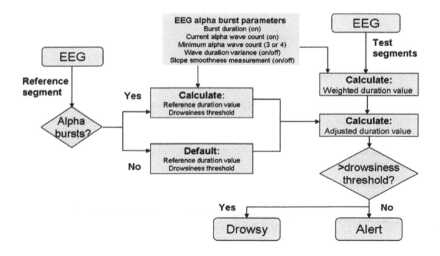

Figure 8.5 Flow diagram of the alpha burst drowsiness detection algorithm.

$$Threshold_{burst} = \frac{K_{burst}}{(1 + e^{-(2\times(Dur_{(ref)}-1.5))})} \tag{8.7}$$

Where K_{burst} is the cut-off parameter of the drowsiness classification for alpha burst analysis, and has a direct impact on sensitivity/specificity. Through analysis, K_{burst} was found to have a range of 3-25, with a default value of 10 which provided the best sensitivity/specificity combination. $Dur_{(ref)}$ is the reference alpha burst duration value, which was calculated in the manner discussed in the following paragraph.

Weighted alpha burst durations ($Dur_{(weight)}$) were calculated from each test segment as in (8.8), using the *burst duration* parameter as the standard parameter, and weighting this value using the other parameters ($Dur_{(ref)}$ was calculated in a similar manner).

$$Dur_{(weight)} = \sum_{i=1}^{n} \prod_{j=1}^{4} Co(param_i(j)) \times burstdur_i \tag{8.8}$$

Where n is the number of bursts within the test interval; *param* is the value of one of the alpha burst parameters other than *burst duration* or *current alpha wave count*; i is the current burst number in the test interval; j is the current alpha burst parameter; *burstdur* is the *burst duration* parameter and Co is the weighting co-efficient.

The weighting co-efficient (Co) of each parameter represented the contribution of the respective parameter. The Co of the *minimum alpha wave count*, $Co_{(wave)}$, was calculated utilising the *current alpha wave count* such that:

$$Co_{(wave)} = 1 + (count_{current} - count_{min}) \times 0.2 \tag{8.9}$$

Where $count_{current}$ is the current alpha wave count and $count_{min}$ is the minimum alpha wave count.

The Co of the other two parameters was assigned a value of $0.5 - 1.5$, which was proportional to the value of the parameter. I.e. if the *slope smoothness measurement* indicated a clear alpha burst (no superimposed noise), this would result in a higher weighting than if the parameter indicated an ambiguous alpha burst with lots of noise.

The weighted alpha burst duration was then adjusted to compensate for the baseline (reference) values as in (8.10), and this adjusted alpha burst duration ($Dur_{(adjusted)}$) was compared to the threshold value calculated from the reference segment of the EEG to derive the drowsiness index value. If the adjusted alpha duration was greater than the threshold value in either the C_4 or O_2 EEG, than the drowsiness index was set to 1, otherwise it was set to 0.

Table 8.4
Top five results of changing the parameter inputs for the alpha burst drowsiness detection algorithm, based on the greatest AROC value

Min alpha wave count	Slope smooth. measure	Wave duration variance	AROC	Sens at 99% Spec	Sens at 98% Spec	Sens at 95% Spec
4	On	On	0.764	0.599	0.678	0.744
4	On	Off	0.763	0.601	0.680	0.742
3	On	Off	0.763	0.590	0.640	0.733
3	On	On	0.762	0.590	0.638	0.745
4	Off	Off	0.758	0.599	0.670	0.746

$$Dur_{(adjusted)} = Dur_{(weight)} - Dur_{(ref)} \qquad (8.10)$$

The *burst duration* and *current alpha wave count* were the consistent parameters that formed the base of the drowsiness detection algorithm, and the other 3 parameters were varied to determine the optimum combination for accurate detection of drowsiness.

Table 8.4 shows the combination of parameters that resulted in the 5 greatest AROC values for the alpha burst drowsiness detection algorithm. For explanation of the method used to validate this algorithm, refer to [38]. The best performing algorithm utilised a *minimum alpha wave count* of 4, and included both the *slope smoothness measurement* and *wave duration variance* parameters as input arguments. The resulting AROC value for this algorithm was 0.764 with a sensitivity/specificity value of 0.742/0.955. This is an increase of 0.045 in the AROC value over the alpha power drowsiness detection algorithm (spectral analysis).

The sensitivity of this algorithm remained ≤ 0.85 even when specificity was < 0.8. This indicated a false negative detection rate of 15% or greater, showing that some of the drowsiness events are being missed by the algorithm. As can be seen by these results, there remains room for improvement in this drowsiness detection. In the next section, the association between drowsiness level and its physiological indicators is tested to find out if it can be strengthened using the secondary signal source of body movements.

8.4.3 Increasing strength of assocciation by adding body movement measurements

From the work by Wierwille and Ellsworth [6], it is known that as drowsiness increases, body movements progress from the normal occasional head, arm or body movement onto restlessness in the very early stages of drowsiness. Then as drowsiness progresses further, periods that are void of any body movements occur, followed by large or abrupt corrective movements. These periods of lack of movement increase in length and frequency with increasing severity of drowsiness and these changes in body movement are the reason that body movement was chosen as the secondary drowsiness indicator to strengthen the association between physiological indicators and drowsiness level.

Movements on the driver's seat were detected using 10 piezoelectric film sensors that were equally distributed in the bottom (5 sensors) and back (5 sensors) sections of the driver's seat. Seat movement magnitude was calculated as peak to peak values over 2s intervals with 1s overlap, and averaged for a 30s period of interest. Changes in movement magnitude were normalised using a reference movement magnitude calculated from an alert baseline at the start of the simulated driving period.

The association between the alpha power and driver drowsiness was reassessed using the seat movement magnitude as an additional indicator of drowsiness. As with before, the adjusted maximum alpha power from C_4 and O_2 were used as the independent variables in the multivariate linear regression model to predict the average drowsiness level of a 30 second period, along with the seat movement magnitudes from each piezoelectric sensor. of 0.033.

Table 8.5 provides a break-down of the regression coefficients, t and p-values of the linear regression, including only the statistically significant variables. These results indicate an increase in the strength of the linear relationship between drowsiness level and the alpha power with body movements ($R^2 = 0.214$), compared to the alpha power without movement data ($R^2 = 0.181$), where there has been an increase in the goodness of fit with this statistical model of 0.033.

Table 8.5

Parameters of the linear regression model for average drowsiness predicted from the maximum relative EEG alpha power and body movement magnitude.

Covariate	Regress. Coeff.	t-value	p-value	R^2
$Max_{adjusted}(C_4)$	1.121 [0.912; 1.331]	10.51	<0.001	
$Max_{adjusted}(O_2)$	2.003 [1.824; 2.183]	21.87	<0.001	
Movement Sensor 1	4.814 [1.883; 7.745]	3.22	0.001	
Movement Sensor 2	8.527 [3.045; 14.008]	3.05	0.002	
Movement Sensor 3	-8.714 [-11.766; -5.662]	-5.6	<0.001	0.214
Movement Sensor 4	5.227 [1.455; 9]	2.72	0.007	
Movement Sensor 5	-31.103 [-36.249; -25.957]	-11.85	<0.001	
Movement Sensor 8	-16.686 [-21.163; -12.209]	-7.31	<0.001	
Movement Sensor 9	14.132 [8.242; 20.022]	4.7	<0.001	
Movement Sensor 10	15.245 [11.886; 18.605]	8.9	<0.001	

These results have indicated a strengthening of association between drowsiness level and its physiological predictors through the combination of body movement and EEG data. It is therefore possible that the addition of body movement indicators in the drowsiness detection algorithms developed earlier would increase the accuracy of these algorithms.

8.5 CONCLUSIONS

This chapter has provided a general overview of the indicators of drowsiness in the vehicle operator, and some current methods of detecting driver drowsiness. The chapter has also given an in-depth example of two approaches to monitoring driver drowsiness using power spectral analysis and pattern recognition in the spontaneous EEG. It was then demonstrated that the association between indicators and drowsiness level can be increased through the use of an additional indicator such as body movement magnitude.

8.6 PROBLEMS AND SOLUTIONS

Problem 1
Using Matlab generate a 30 second length of test signal sampled at 256Hz, that resembles the drowsy EEG. This signal should contain a mixture of waxing and waning beta activities (use 16, 23 and 31Hz) and at least 3 bursts of alpha activity (use 10Hz, with bursts lasting 0.7, 1.4 and 1.1 seconds). Separately display the whole signal, and 8 seconds of signal surrounding one of the alpha bursts.

Solution 1
```
%step 1 - make a time function 30 seconds in length with a
sampling rate of 256Hz
samp = 256;
t = 1/samp:1/samp:30;

%step 2 - make the original beta rhythms, using decreasing
amplitude for increasing
%frequency, as would be expected in an EEG signal
beta1 = 0.4*sin(2*pi*16*t);
beta2 = 0.3*sin(2*pi*23*t);
beta3 = 0.2*sin(2*pi*31*t);

%step 3 - make masks for the beta signals, that makes them
resemble a signal that is waxing
%and waning at different rates
mask1 = 0.1*sin(2*pi*0.4*t)+1;
mask2 = 0.2*sin(2*pi*0.2*t)+1;
mask3 = 0.1*sin(2*pi*0.3*t)+1;
```

```
beta1 = beta1.*mask1;
beta2 = beta2.*mask2;
beta3 = beta3.*mask3;

%step 3 - make the three alpha bursts
alpha1 = sin(2*pi*10*t(1:180)); %0.7 seconds in length
alpha2 = sin(2*pi*10*t(1:359)); %1.4 seconds in length
alpha3 = sin(2*pi*10*t(1:282)); %1.1 seconds in length

%step 4 - apply tukey window to alpha bursts to taper the edges of
the bursts
wind1 = tukeywin(180,0.3);
wind2 = tukeywin(359,0.3);
wind3 = tukeywin(282,0.3);
alpha1 = alpha1.*wind1';
alpha2 = alpha2.*wind2';
alpha3 = alpha3.*wind3';

%step 5 - put alpha bursts into a single array worth 30 seconds in
length
alpha = zeros([1 30*samp]);
alpha(310:489) = alpha1;
alpha(3465:3823) = alpha2;
alpha(5862:6143) = alpha3;

%step 6 - combine the 3 signals together and plot this signal,
adding a title, x-axis label
%and setting the y limits large enough to clearly see the signal
y = beta1 + beta2 + beta3 + alpha;
figure(1)
plot(t(:),y(:));
title('\fontsize{14}Test signal of mixed \beta with bursts of
\alpha');
xlabel('\fontsize{12}time (seconds)');
ylim([-15 15]);

%then plot this signal concentrating on the 10 to 18 seconds
period
figure(2)
plot(t(:),y(:));
title('\fontsize{14}8 seconds of test signal surrounding one
\alpha burst');
xlabel('\fontsize{12}time (seconds)');
ylim([-15 15]);
xlim([10 18]);
```

Figure 8.6 is the output of the MatLab program for Problem 1.

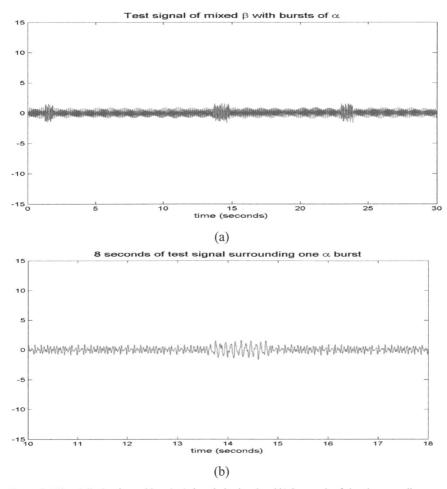

Figure 8.6 Signal display for problem 1, a) the whole signal and b) 8 seconds of signal surrounding an alpha burst.

Problem 2
Create the Power Spectral Density (PSD) arrays matrix of the test signal generated in problem 1, using the FFT function. For this, a moving window of 1 second length and 0.75 second overlap should be used. Display matrix of PSD arrays on a 3D plot over time, restricting the frequency displayed to within 1 – 35Hz.

Solution 2

```
%The PSD is found using the variables already created when solving
the 1st problem. Using a
```

%while loop allows for a moving window and the creation of a PSD array matrix, based on 1
%second segment lengths. A few variables are defined prior to the loop.

```
a = 0;
PSD = [];   %assigning an empty variable to build the array matrix
n = 1*samp;   %define the length of the test segment and the FFT
function
```

%To reduce the noise created at the edges of the fixed length signal, the signal needs to
%be "windowed" so that the edges taper to zero amplitude. In this example, a hamming window
%is utilised

```
wind = hamming(n);

while a <= (length(y)-n)
    % define the test segment
    seg = y((a+1):(a+n));
```

%When working with biological signals, it is best to remove the DC component of the
```
    %signal, so that the signal has a zero mean
    seg = seg(:) - mean(seg(:));
```

```
    %apply the windowing function
    seg = seg.*wind;
```

```
    %perform the FFT function
    Y = fft(seg,n);
```

```
    %the power is then found as follows
    Py = Y.*conj(Y)/512;
```

%add this array to the PSD array matrix, so that each row in this matrix holds the
```
    %array of one iteration of the FFT function
    PSD = cat(1,PSD,Py');
```

%move the window forward by 0.25 seconds to result in a 0.75 second overlap
```
    a = a + (0.25*samp);
end
```

%only the first half of the power array is meaningful, the rest of the data points are

```
%redundant.  To  find  the  frequencies  for  plotting  the  power
spectral density array, the
%sample rate and the number of points in the FFT need to be used.
f = samp*(0:n/2)/n;

%make a 3D plot of the PSD arrays over time, with the range of 1 -
35Hz, adding a title and
%x-axis label. The frequencies makes the x-axis, the times (or
iterations) makes the y-axis
%and the power makes the z-axis.

%First, make the appropriate sized times matrix
times = zeros([size(PSD,1) 35]);
x = 0:0.25:29;
x = x';
a = 1;
while a <= 35
    times(:,a) = x;
    a = a + 1;
end

%then plot using the plot3 function
figure(3)
plot3(f(2:36),times,PSD(:,2:36),'b');
title('\fontsize{14}Time  course  of  the  PSD  arrays  of  the  test
signal');
xlabel('\fontsize{12}frequency (Hz)');
ylabel('\fontsize{12}start time of iterations');
zlabel('\fontsize{12}power');
```

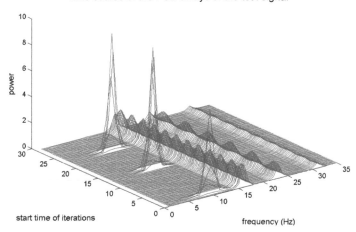

Figure 8.7 Power Spectral Density plot for problem 2.

Figure 8.7 is a power spectral density plot for problem 2.

Problem 3
Find the maximum relative alpha power (Max_{alpha}) for the 30 seconds of test signal as described in section 8.4.2.1.

Solution 3
```
%The maximum relative alpha power was found as the maximum of the
relative alpha values,
%which were averaged over consecutive 2.25 second segments.

%step 1 - find the relative alpha power values for each iteration
of the FFT
RelPow = zeros([1 size(PSD,1)]);
a = 1;

while a <= size(PSD,1)
    RelPow(a) = sum(PSD(a,9:14))/sum(PSD(a,2:36));
    a = a + 1;
end

RawMax = max(RelPow);

%step 2 - find relative power averages over 2.25 seconds (6
iterations)
AvePow = zeros([1 20]);
a = 1;
b = 1;

while a < 20
    AvePow(a) = mean(RelPow(b:b+5));
    a = a + 1;
    b = b + 6;
end

AvePow(20) = mean(RelPow(b:end));

%step 3 - find maximum alpha power from the averaged relative
powers
MaxAlpha = max(AvePow);
```

The maximum relative alpha power prior to averaging to remove spurious data is 80.343%

The Max_{alpha} value found as described in section 8.4.2.1 is 56.464%

References

[1] R. C. Coetzer and G. P. Hancke, "Driver fatigue detection : A survey," in AFRICON,23-25 Sept, 2009, pp. 1-6.

[2] M. Rimini-Doering, T. Altmueller, U. Ladstaetter, and M. Rossmeier, "Effects of Lane Departure Warning on Drowsy Drivers' Performance and State in a Simulator," in First International Driving Symposium on Human Factors in Driver Assessment, Training and Vehicle Design2001, pp. 88-95.

[3] Transport Accident Commission, "Reducing Fatigue - a Case Study," Victoria.

[4] C. L. Baldwin and J. F. May, "Driver fatigue: The importance of identifying causal factors of fatigue when considering detection and countermeasure technologies," Transportation Research Part F: Traffic Psychology and Behaviour, vol. 12, pp. 218-224, 2009.

[5] K. Dobbie, "Fatigue Related Crashes: An Analysis of Faitgue-related Crahes on Australian Roads using an Operational Definition of Fatigue," Australian Transport Safety Bureau Road Safety Research Report OR 23, 2002.

[6] W. W. Wierwille and L. A. Ellsworth, "Evaluation of driver drowsiness by trained raters," Accident Analysis & Prevention, vol. 26, pp. 571-581, 1994.

[7] E. Vural, M. Cetin, A. Ercil, G. Littlewort, M. Bartlett, and J. Movellan, "Drowsy driver detection through facial movement analysis," Lecture Notes in Computer Science, vol. 4796, pp. 6-18, 2007.

[8] S. H. Fairclough and R. Graham, "Impairment of Driving Performance Caused by Sleep Deprivation or Alcohol: A Comparative Study," Human Factors: The Journal of the Human Factors and Ergonomics Society, vol. 41, pp. 118-128, March 1 1999.

[9] M. G. Khan, "Basic Concepts " in Rapid ECG Interpretation, C. P. Cannon, Ed.: Humana Press, 2008, pp. 1-24.

[10] M. R. Volow and C. W. Erwin, "Heart Rate Variability Correlates of Spontaneous Drowsiness Onset," SAE Technical Paper 730124, 1973.

[11] I. Takahashi and K. Yokoyama, "Development of a feedback stimulation for drowsy driver using heartbeat rhythms," in Annual International Conference of the IEEE Engineering in Medicine and Biology Society, EMBC, Boston, MA2011, pp. 4153-4158.

[12] C. Iber, S. Ancoli-Israel, A. L. Chesson, and S. F. Quan, The AASM Manual for the Scoring of Sleep and Associated Events: Rules, Terminology and Technical Specifications. Westchester, IL: American Acadamy of Sleep Medicine, 2007.

[13] L. N. Boyle, J. Tippin, A. Paul, and M. Rizzo, "Driver performance in the moments surrounding a microsleep," Transportation Research Part F: Traffic Psychology and Behaviour, vol. 11, pp. 126-136, 2008.

[14] Y. Tran, A. Craig, N. Wijesuriya, and H. Nguyen, "Improving classification rates for use in fatigue countermeasure devices using brain activity," in Annual International Conference of the IEEE Engineering in Medicine and

Biology Society (EMBC), 2010 Aug. 31 -Sept. 4 2010, pp. 4460-4463.

[15] J. R. Hughes, EEG in clinical practice. Boston: Butterworth-Heinemann, 1994.

[16] J. Santamaria and K. H. Chiappa, The EEG of drowsiness. New York: Demos Publications, 1987.

[17] J. Cantero, M. Atienza, and R. Salas, "Human alpha oscillations in wakefulness, drowsiness period, and REM sleep: different electroencephalographic phenomena within the alpha band," Neurophysiologie Clinique, vol. 32, pp. 54-71., Jan 2002.

[18] L. Torsvall and T. Akerstedt, "Sleepiness on the job: continuously measured EEG changes in train drivers," Electroencephalography and Clinical Neurophysiology, vol. 66, pp. 502-511, 1987.

[19] Q. Wang, J. Yang, M. Ren, and Y. Zheng, "Driver Fatigue Detection: A Survey," in The Sixth World Congress on Intelligent Control and Automation, WCICA 2006. 2006, pp. 8587-8591.

[20] Y. Dong, Z. Hu, K. Uchimura, and N. Murayama, "Driver Inattention Monitoring System for Intelligent Vehicles: A Review," IEEE Transactions on Intelligent Transportation Systems, vol. 12, pp. 596-614, 2010.

[21] S. Bhuiyan, "Driver Assistance Systems to Rate Drowsiness: A Preliminary Study," in New Advances in Intelligent Decision Technologies. vol. 199, K. Nakamatsu, G. Phillips-Wren, L. Jain, R. Howlett, and S. Bhuiyan, Eds. Berlin / Heidelberg: Springer 2009, pp. 415-425.

[22] R. Sayed and A. Eskandarian, "Unobtrusive drowsiness detection by neural network learning of driver steering," Proceedings of the Institution of Mechanical Engineers, Part D: Journal of Automobile Engineering, vol. 215, pp. 969-975, September 1 2001.

[23] L. Fletcher, N. Apostoloff, L. Petersson, and A. Zelinsky, "Vision in and out of vehicles," Intelligent Systems, vol. 18, pp. 12-17, 2003.

[24] W. S. Wijesoma, K. R. S. Kodagoda, and A. P. Balasuriya, "Road-boundary detection and tracking using ladar sensing," IEEE Transactions on Robotics and Automation, vol. 20, pp. 456-464, 2004.

[25] W. W. Wierwille, M. G. Lewin, and R. J. I. Fairbanks, "Final Report: Research on Vehicle-Based Driver Status/Performance Monitoring; Part 1," Virginia Polytechnic Institute and State University, Virginia 1996.

[26] D. Liu, P. Sun, Y. Xiao, and Y. Yin, "Drowsiness detection based on eyelid movement," in 2nd International Workshop on Education Technology and Computer Science, Wuhan, Hubei, China2010, pp. 49-52.

[27] T. Danisman, I. M. Bilasco, C. Djeraba, and N. Ihaddadene, "Drowsy driver detection system using eye blink patterns," in 2010 International Conference on Machine and Web Intelligence (ICMWI),3-5 Oct. 2010, 2010, pp. 230-233.

[28] Y.-S. Wu, T.-W. Lee, Q.-Z. Wu, and H.-S. Liu, "An Eye State Recognition Method for Drowsiness Detection," in IEEE 71st Vehicular Technology Conference (VTC 2010-Spring) 16-19 May 2010, 2010, pp. 1-5.

[29] S. Hu, R. L. Bowlds, Y. Gu, and X. Yu, "Pulse wave sensor for non-intrusive driver's drowsiness detection," in Annual International Conference of the IEEE Engineering in Medicine and Biology Society, 2009. (EMBC). ,3-6 Sept. 2009, 2009, pp. 2312-2315.

[30] M. V. Ramesh, A. K. Nair, and A. T. Kunnath, "Intelligent Steering Wheel Sensor Network for Real-Time Monitoring and Detection of Driver Drowsiness," International Journal of Computer Science and Security, vol. 1, pp. 1-9, 2011.

[31] A. Picot, S. Charbonnier, and A. Caplier, "On-line automatic detection of driver drowsiness using a single electroencephalographic channel," in 30th Annual International Conference of the IEEE Engineering in Medicine and Biology Society, EMBS 2008. ,20-25 Aug. 2008, 2008, pp. 3864-3867.

[32] M. Simon, E. A. Schmidt, W. E. Kincses, M. Fritzsche, A. Bruns, C. Aufmuth, M. Bogdan, W. Rosenstiel, and M. Schrauf, "EEG alpha spindle measures as indicators of driver fatigue under real traffic conditions," Clinical Neurophysiology, vol. 122, pp. 1168-1178, 2011.

[33] A. Sathyanarayana, S. Nageswaren, H. Ghasemzadeh, R. Jafari, and J. H. L. Hansen, "Body sensor networks for driver distraction identification," in IEEE International Conference on Vehicular Electronics and Safety, Piscataway, NJ, USA2008, pp. 120-5.

[34] D. Lee, S. Oh, S. Heo, and M. Hahn, "Drowsy driving detection based on the driver's head movement using infrared sensors," in International Symposium on Universal Communication, Piscataway, NJ, USA2008, pp. 231-6.

[35] S. Pritchett, E. Zilberg, Z. M. Xu, M. Karrar, S. Lal, and D. Burton, "Strengthening Association between Driver Drowsiness and its Physiological Predictors by Combining EEG with Measures of Body Movement," in 6th International Conference on Broadband Communications & Biomedical Applications (IB2COM 2011) Melbourne, 2011.

[36] G. R. Kecklund and T. R. Åkerstedt, "Sleepiness in long distance truck driving: an ambulatory EEG study of night driving," Ergonomics, vol. 36, pp. 1007-1017, 09/01, 2011/10/17 1993.

[37] M. Karrar, E. Zilberg, Z. M. Xu, D. Burton, and S. Lal, "Detection of Driver Drowsiness using EEG Alpha Wave Bursts - Comparing Accuracy of Morphological and Spectral Algorithms," in International Conference on Fatigue Management in Transportation Operations: A Framework for Progress, Boston, USA, March 24-26, 2009.

[38] S. Pritchett, E. Zilberg, Z. M. Xu, M. Karrar, D. Burton, and S. Lal, "Comparing the Accuracy of Two Algorithms for Detecting Driver Drowsiness - Single Source (EEG) and Hybrid (EEG and Body Movement)," in 6th International Conference on Broadband Communications & Biomedical Applications (IB2COM 2011) Melbourne, 2011.

[39] E. Brigham and R.E. Morrow, "The Fast Fourier Transform," Spectrum, IEEE, vol. 4, pp. 63-70, 1967.

Chapter 9.

Cortical Width Measurement Based on Panoramic Radiographs Using Computer-Aided System

Przemyslaw Mackowiak[1], Elżbieta Kaczmarek[2], and Tomasz Kulczyk[3]

[1]*Poznan University of Technology, Poland,*
[2]*Department of Bioinformatics and Medical Biology, Poznan, University of Medical Sciences*
[3]*Department of Biomaterials and Experimental Dentistry, Poznań University of Medical Sciences*
przemyslaw.mackowiak@gmail.com; elka@amp.edu.pl;
tomasz.kulczyk@gmail.com

9.1 PROBLEM ANALYSIS

The chapter describes a problem of the cortical width measurement on a dental panoramic radiograph. The cortical width may potentially be associated with recognition of osteoporosis in the case of postmenopausal women. The measurement is usually carried out by an expert (like a radiologist). They use a ruler and a pen to compute distance between two points of the mandibular on a dental image. This method is unfortunately time-consuming.

The following chapter is about the cortical width determination using computer-aided system. The proposed algorithm is based on separate extraction of the outer and inner margin of the mandible.

Furthermore, we evaluate the performance of the proposed algorithm. The comparison was carried out based on the hand-made measurement by two radiologists. Thirty four dental panoramic radiographs of healthy and osteoporotic individuals were taken into account.

A dental panoramic radiograph shows an excellent view of mandible with

Bio-Informatic Systems, Processing and Applications, 169-190,

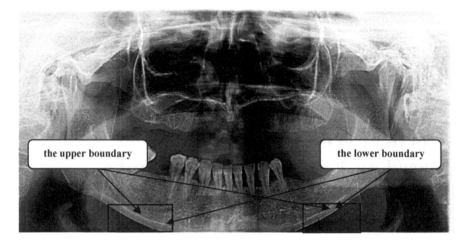

the upper boundary

the lower boundary

Figure 9.1 The panoramic radiograph with two black rectangles pointing at the regions of interest. White paths are the upper and lower boundary of mandibular bone.

clear details of cortical bone. Dental panoramic radiographs are used extensively in dental practice all over the world. A sample image was shown in Figure 9.1. What is the main motivation of the cortical width (CW) computing? Some authors conclude that if the cortical width is less than 3mm, an individual should be advised to seek further osteoporosis investigation [5]. This makes the development of the computer aided system an interesting idea. It could help to decrease the measurement time carried out by a specialist.

Unfortunately, the cortical width is usually computed by hand. The input image is a dental panoramic radiograph. A radiologist draws the tangent to the lower cortical bone boundary as it was illustrated in Figure 9.2(b). Then, the radiologist draws a second line that crosses the mental foramen and is perpendicular to the tangent (see Figure 9.2(c)). The measurement is carried out along the perpendicular line. The cortical width is the Euclidian distance between two points which lie on the perpendicular line. One of them is the intersection point (A), second one (B) lies on the outer margin (see Figure 9.2(d)).

At this point, it is important to include herein the information about osteoporosis. It is characterized by low bone mass density (BMD) and deterioration of microarchitecture of bone. It is recognized as one of causes of morbidity and mortality in the case of postmenopausal women [1]. Osteoporosis can be recognized using the double-energy X-ray absorptiometry (DEXA) performed in the region L1-L4 of spinal column or a hip. This method is expensive and requires appropriate equipment. The examinations cannot be carried out on large population due to low accessibility. Therefore scientists are looking for cheaper and more accessible methods. These methods must not necessarily be used to make a full diagnosis of the osteoporosis. They should be helpful for screening larger populations. Medical scientists think that dental

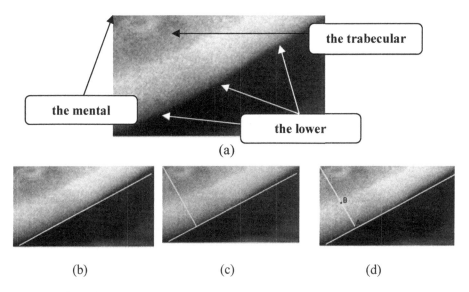

Figure 9.2 The cortical width measurement.

panoramic radiographs are of use in the osteoporosis investigation. Some anatomical and pathological structures can be found on those radiographs. Information provided by panoramic images can be used to establish a proper diagnosis, to propose a treatment plan and to evaluate the results of the treatment. Scientists search for features, which are correlated with low BMD. Both morphometric and morphological features are sought. One of them is the cortical width. Some scientists reveal correlation between CW and BMD results. Last works demonstrate that cortical width measurement may be used in identifying women with low BMD level [2].

9.2 METHOD OUTLINE

The description of the computer-aided system for the cortical width measurement is the main goal of the chapter. The outline of the method is presented below. As a first step, the region of interest is determined. Then, extraction of the lower cortical bone boundary (the inner margin) is performed. The next step is the upper boundary (the outer margin) determination. Finally, the cortical width is measured. The above steps are described in the further sections of the present chapter. Figure 9.3 illustrates the outline of the method proposed.

9.3 REGION OF INTEREST DETERMINATION

An expert (like a radiologist) cut the rectangular region of interest with respect to the mental foramen to include as large mandibular structure part as possible. The black rectangles were marked on the Figure 9.1. They indicate the regions of interest, where the cortical width is computed. The regions of interest represent bone morphological properties sufficiently and have smaller size than an original image, which results in decreasing of the computation time. The area received in this way contains both the mandibular and the trabecular bone. The sample region of interest was shown in Figure 9.2(a).

9.4 LOWER BOUNDARY EXTRACTION

The lower boundary is extracted based on two steps. First the gradient image is computed. Then the path determination method is used.

The lower cortical bone boundary was shown in Figure 9.2(a). It lies in the region of the high contrast. We can try to detect this boundary using some edge detection methods. One of the most common methods is gradient computing followed by thresholding. Sobel, Prewitt or Roberts mask can be used. We should consider using some noise attenuation technique like median filtering or average filter with different size of masks (3x3, 5x5, ...). The greater mask, the more attenuated noise. The gradient image is computed based on the equations given below

$$G(m,n) = \sqrt{\left|I(m+1,n) - I(m,n)\right|^2 + \left|I(m,n+1) - I(m,n)\right|^2}$$

$$G(m,n) = max\left(\left\|I(m+1,n) - I(m,n)\right\|, \left|I(m,n+1) - I(m,n)\right\|\right) \qquad (9.1)$$

where (m,n) are pair of coordinates, G(m,n) determines the gradient module for coordinates (m,n), I(m,n) is the input image pixel whose coordinates are (m,n). The thresholding is carried out to receive the edge map. The main problem lies in the appropriate selection of threshold. It can have global (global thresholding) or local (local thresholding) value. The thresholding does not ensure that only the lower cortical boundary is received. Unfortunately, many additional objects appear. These objects are associated with the upper mandibular bone boundary or with trabecular bone. The object which corresponds to the lower boundary has not one pixel thickness. The perimeter is computed to receive the lower object border. Its lower part can be approximately treated as the lower cortical bone boundary. The method, has however many disadvantages. The threshold selection is crucial and leads to the appropriate edge map. The contrast varies along the lower boundary. This makes that the lower boundary received can be broken

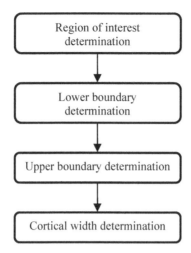

Figure 9.3 Block diagram of the proposed algorithm.

(discontinuity) when the global threshold is used. The above observation suggests using of the adaptive threshold instead of the global one. The algorithm does not assure that the lower boundary received is continuous along the real boundary. The method does not seem appropriate in the present application.

Canny method or Laplacian of Gaussian (LoG) and zero-cross searching seem to be more suitable approaches. The problem arises with the choice of the appropriate kernel size. The settings of thresholds in Canny method are crucial and their experimental selection for each image is time consuming. The above mentioned algorithms do not ensure of the continuity of the lower boundary either. They unfortunately extract more than one edge.

Taking into account the listed facts, we propose an algorithm which generates single edge only. Gradient computing and path determination are two parts that describe the method. They are explained in sections 9.4.1 and 9.4.2.

9.4.1   *Gradient Computing*

The gradient can be computed in different ways. Two approaches were mentioned in the previous section. They base on the nearest neighbors of pixel of interest. We propose gradient computing technique based on the idea presented in [6]. Out technique is slightly different because three directional derivatives are taken into account. This is a multi-scale approach which uses the different kernel size to adjust to the spatial edge size. Figure 9.4 presents the block diagram which describes the gradient computing method. $G_{2k}45$, $G_{2k}90$, $G_{2k}135$ are blocks that compute the directional derivative for the angles $45°$, $90°$ and $135°$ respectively (derivatives for the remaining angles are not computed). The choice of the angles is the result of the spatial orientation of the lower mandibular bone boundary in

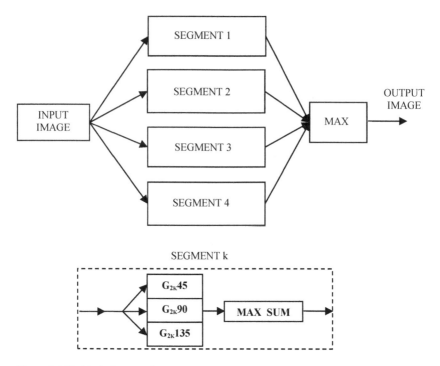

Figure 9.4 The block diagram of the gradient computing method.

the region of interest. The highest values of gradient are for derivatives computed for the angle equal to either 45° or 90° for the right side image, and for derivatives computed for the angle equal to 90° or 135° for the left side. The greater index k (k={1,2,3,4}), the greater kernel size. The kernels are created using the modified derivative Sobel operator [8]. To explain the above mentioned operator more deeply we present the process of kernel creating for the angle equal 90°. The partial derivative in y direction (90°) can be approximated by standard Sobel operator [3x3] presented in Figure 9.5. We can express the operator as a polynomial function using a 'i-j' coordinate system. Therefore, the operator can be represented by following terms:

$i^2+2i^2j+i^2j^2-1-2j-j^2=i^2\cdot(1+2j+j^2)-(j+1)^2=i^2\cdot(j+1)^2-(j+1)^2=(i^2-1)\cdot(j+1)^2$
$=(i-1)\cdot(i+1)\cdot(j+1)^2$.

This example is helpful to create a hierarchy of y derivative Sobel operator (90°):

$$\frac{\partial}{\partial y} = (j+1)^n \cdot (i+1)^{n-1} \cdot (i-1), \ k = 2\cdot n, \text{ for n} > 0 \tag{9.2}$$

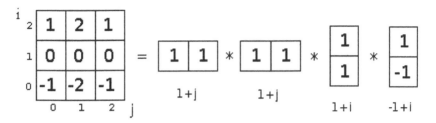

Figure 9.5 The decomposition of the standard Sobel operator for the angle equal 90°.

The standard Sobel operator is generated for k equal 1. In similar way the modified Sobel operator is created for angles equal 45° and 135°. The derivatives computed for different angles for the same kernel size are inputs of MAX SUM block. This block sums the two greatest values and its output is the input of MAX lock. The MAX block selects the greatest value which is treated as the final

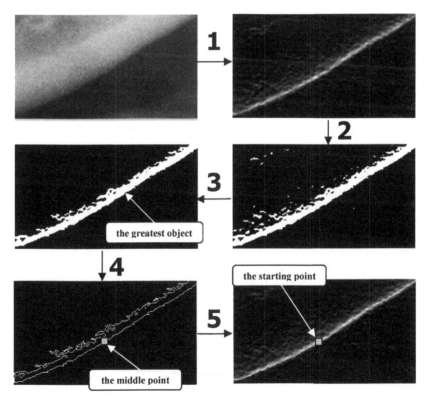

Figure 9.6 The lower boundary extraction. 1 - gradient computing, 2 – thresholding, 3 – greatest object selection, 4 – middle point selection, 5 – the gradient image with the white path that corresponds to the lower boundary extracted).

gradient value at the pixel of interest.

It is important to emphasize that if we know which image side is processed, the block diagram presented in Figure 9.4 can be greatly reduced. In such situation only two derivatives are computed (for angles equal 135° and 90° for the left image side or for angles equal 45° and 90° for the right image side). Then, the MAX_SUM block is replaced by simple SUM block that sums two numbers. The way in which a specialist cuts the region of interest can point at the appropriate image side.

In this way the gradient image is created. It is further used as the input image for the path determination method.

9.4.2   Path Determination

This is the second and the last step of the extraction of the lower cortical bone. At the beginning some starting point has to be selected. It is located in the region of the high gradient values (here called by "mountain region", due to its similarity to mountain peaks) which corresponds to the area where the boundary lies. The sample starting point was shown in Figure 9.6. To exclude the problem of the manual selection of the starting point, we propose a method which determines it automatically. The binary image is created based on the global thresholding. We assume that the cutoff threshold equals to 90th percentile level. Thanks to that, 10 percent of the highest value pixels are received. The greatest gradient values are located in the region of the lower boundary. The object whose surface is the greatest is taken under analysis. A middle point of its lower boundary is selected. Its location is corrected vertically to the nearest brightness maximum. The relocated point is the starting point. Since the starting point is given, the next step which can be compared to the "walking in the mountains" is carried out. Its result is a path which corresponds to the lower boundary. The block diagram of the algorithm was shown in Figure 9.7. In the method six areas of the search around the reference point (the reference point is the starting point at the beginning) are defined. They are called A, B, C, D, E, F and were illustrated in Figure 9.8. The regions contain nine pixels. The weights associated with each region were selected by an expert and presented in Table 9.1. The motivation was to favor some directions of the search where the pixels belonging to the boundary should lie. At the beginning only the regions D, E, F are taken into account. The string

Table 9.1
The weights example

Region	A	B	C	D	E	F
Weights*	1.0	1.0	1.5	1.5	1.5	1.0
Weights**	1.5	1.5	1.0	1.0	1.0	1.5

*- for left image side, **- for right image side

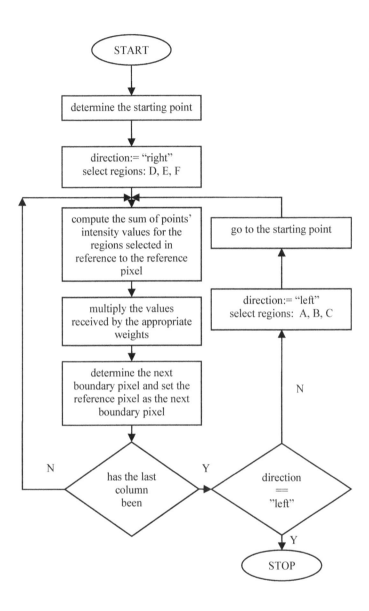

Figure 9.7 The block diagram of the path determination method.

"right" is assigned to the helpful variable direction that determines roughly the direction of the extraction. A sum of points' intensity values is computed separately for each region. For example the sum of points' intensity values of the region D is the sum of intensity values of points that belong to the region D. Thanks to that three values are received which are multiplied further by the appropriate weights. The region associated with the highest value is selected. The boundary pixel is the pixel associated with the chosen area. The boundary pixel is pixel N2 when region D is selected, pixel N4 when region E is selected, pixel N6 when region F is selected. The selected pixel becomes the new reference pixel.

If the last column of an image is reached, the string "left" is assigned to the variable direction and the process starts from the beginning. Then, only the areas A, B, C are taken into account. The boundary is extracted in the opposite direction starting from the starting point. The set of selected reference pixels correspond to the lower mandibular boundary.

9.5 UPPER BOUNDARY EXTRACTION

The example upper boundary was presented in Figure 9.9. It lies in the weak contrast region. We assume that the boundary points correspond to places where second derivative has zero value. The inflection points need to be extracted because they correspond to the points which belong to the upper bone boundary. The method we want to describe is two-phase approach. The maximums estimation is carried out first, then the inflection estimation is done. The former operation allows for more accurate extracting the inflection points.

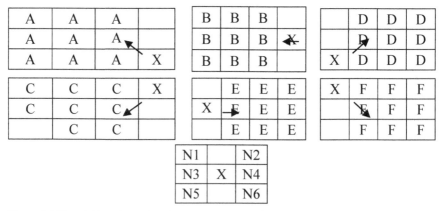

Figure 9.8 The neighbors (N1-N6) and areas (A-F) around the pixel of interest.

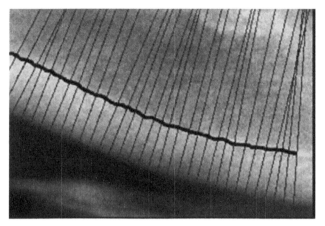

Figure 9.9 The black path corresponds to the upper boundary extracted by a radiologist. The normal to the lower boundary are profiles (they are normal to the lower boundary).

9.5.1 Maximums Estimation

The maximums estimation is based on the lower boundary extracted in the section 9.3. Every k-th pixel of the lower boundary is selected (for example k=4). Next, the profiles are drawn. They cross the chosen pixels and are normal to the lower boundary at the specified pixels. The profiles are illustrated in Figure 9.9. The profile can be understood as the function of pixel brightness depending on the location. Figure 9.10 shows a sample profile. Each profile is filtered with the use of two-dimensional Gaussian filter to decrease noise influence. Different kernels are used depending on the parameter σ. It corresponds to the standard deviation of the Gaussian function ($\sigma=\{1.0,1.5,2.0,2.5,3.0,3.5,4.0,4.5\}$). Eight denoised profiles correspond to one noised profile. The location of the first brightness maximum of denoised profiles is found (see Figure 9.11(b)). They create eight element series. Then, one maximum is estimated based on the set of eight maximums. We introduce supporting parameters m and p. The former defines the length of series, the latter the difference. First m-elements series whose elements do not differ by more than p is found. The estimated maximum is the value of the first series element. We will consider an example to explain the method deeply. Let the maximums found be kept in table [30, 31, 45, 46, 46, 46, 47, 47]. Each cell contains the number which corresponds to some point number of the profile. Each number is associated with the first maximum found for the denoised profiles (e.g., number 31 corresponds to the first maximum found for the denoised profile ($\sigma=1.5$)). Let m=3, p=5. The first three elements series whose elements do not differ by more than 5 is [45, 46, 46]. The maximum estimation is the number 45.

Some maximums can be estimated wrong. This is why the one-dimensional median filter is used to correct their locations. Each filtered number is moved to the nearest brightness maximum. Let us consider an example to explain the method more clearly. Table [... 45 34 46 47 48 43 ...] contains numbers which correspond to the maximum estimated for the successive profiles (e.g., number 46 denotes the maximum location of some profile). Let us assume that the value 34 was determined improperly. The numbers are filtered using the median filter (windows span equal to 3). In this case the result is [… 45 **45** 46 47 47 47…]. It is clear that the highlighted value underlined was changed distinctly in comparison to the initial value. Next, the values are moved in the direction of the nearest brightness maximum. The result is [… 45 44 46 47 48 48 …] if the nearest brightness maximums are […45 44 46 47 48 48 …].

The method is used for each profile. We can conclude that the result of the maximum estimation method is set of numbers which correspond to the maximum location for successive profiles. The sample set was presented in Figure 9.12(a) as black bars.

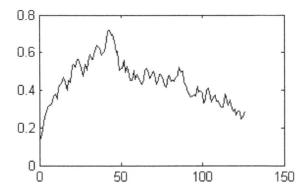

Figure 9.10 The sample profile, x-axis corresponds to the pixel number, y-axis corresponds to the pixel brightness.

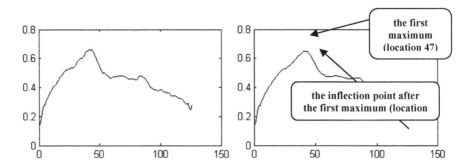

Figure 9.11 Denoised profiles. A - σ =1.0, B - σ =1.5.

9.5.2   Inflection Points Estimation

The inflection points estimation is necessary to point the boundary pixels precisely. One inflection point is associated with one profile. The estimated inflection points correspond to the points that belong to the upper mandibular boundary. For each profile the first inflection point after the maximum estimated in the section 9.4.1 is determined. The estimated inflection points can be noisy, therefore a correction of their location may be needed. We use the median filtering. Then, each filtered point is moved to the nearest inflection point (like in maximums estimation section). Figure 9.12 (b) illustrates the black bars whose length corresponds to the location of the estimated inflection points.

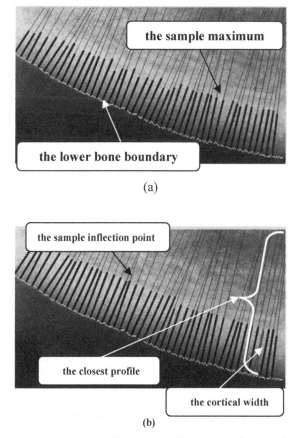

Figure 9.12 The maximums estimation (without median filtering) and inflection points estimation.

9.6 CORTICAL WIDTH MEASUREMENT

The cortical width is measured along some profile. From the set of profiles the profile closest (in the Euclidian distance sense) to the mental foramen is selected and a measurement along this profile is conducted. The profile crosses the lower and upper boundary at two points. The cortical width is the Euclidian distance between them. The gray segment in Figure 9.12 (b) corresponds to the cortical width.

9.7 EXPERIMENTS AND RESULTS

Thirty four dental panoramic radiographs were provided by the Section of Dental Radiology, Department of Biomaterials and Experimental Dentistry, Poznań University of Medical Sciences. The digital images have been taken using the CRANEX TOME dental panoramic unit and a DIGORA PCT PSP DIGITAL SCANNER. The resolution of panoramic images was 300dpi (1pixel corresponds to approximately 0.1mm). The radiologist divided the set of the images into noise free images (seventeen) and cluttered images (seventeen). An image was classified by a specialist as the cluttered image, if the region of interest contained structures which do not appear in a properly acquired image. The improper patient's head setting during the image acquisition process was the main source of the occurrence of the undesired structures. 68 regions of interest (ROI) were taken into account during measurements. Each ROI was presented to two experienced radiologists. They performed manual measurements of cortical width. A comparative analysis was done between the measurement of the cortical width took by program and the measurement of the cortical width took by the radiologists. Both the noise free images and the cluttered images were taken into consideration. The results were expressed by the Bland-Altman plot in Figure 9.13. The x axis corresponds to the average of two measurements (measurements carried out by two radiologists or by one radiologist and program), y axis corresponds to the difference between measurements.

Additionally, the radiologists drew the upper and lower boundary (the reference boundaries). An area around the reference boundary was created to examine the performance of the algorithm. The area is restricted by two curves translated with respect to the reference boundary by d pixels in vertical direction. The example area was shown in Figure 9.14. The total number of pixels generated by the program that lie in the area is computed for different d values. The received value is divided by the total number of points that belong to the reference boundary (the result is expressed in percentages). The computations were carried out for the upper and lower boundary. The results were averaged for all images and were shown in Figure 9.15.

The values of the parameters used were following. Every forth pixel of the lower boundary was selected during the maximums estimation. The weights' values in the case of the lower boundary were the same as in Table 9.1. The m and p parameters were equal to 3 and 5 respectively. The window span was 13 (median filtering).

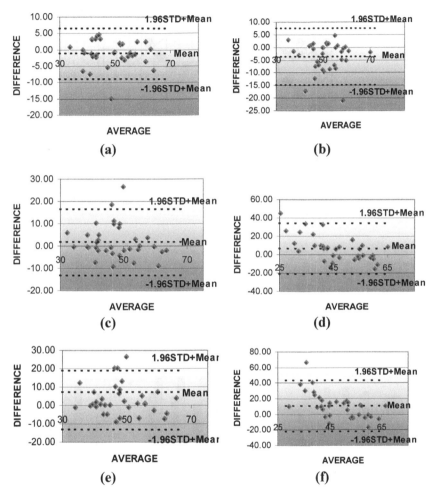

Figure 9.13 The cortical width measurement described by the Bland-Altman plot (STD is the standard deviation of the differences, Mean is the mean value of the differences). (a) - the noise free images (radiologist1-radiologist2), (b) - the cluttered images (radiologist1-radiologist2), (c) - the noise free images (radiologist1-program), (d) - the cluttered images (radiologist1-program), (e) - the noise free images (radiologist2-program), (f) - the cluttered images (radiologist2-program).

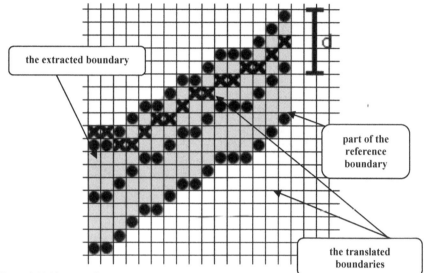

Figure 9.14 The example area (gray color) around the reference boundary for d equal 4. The line drawn with crosses presents the possible boundary generated by program. About 88% (14/16) of the extracted boundary points lie in the defined area.

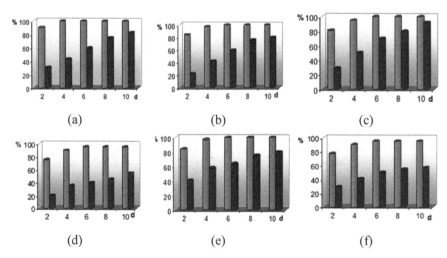

Figure 9.15 The results of the boundary extracted. (a) - the noise free images (radiologist1-radiologist2), (b) - the cluttered images (radiologist1-radiologist2), (c) - the noise free images (radiologist1-program), (d) - the cluttered images (radiologist1-program), (e) - the noise free images (radiologist2-program), (f) - the cluttered images (radiologist2-program). The gray bars correspond to the lower boundary, the black bars correspond to the upper boundary.

9.8 DISCUSSION

The greatest problem in the cortical width measurement and boundaries extraction issue is the lack of the absolute "ground truth" (boundaries shape, cortical width measured). Therefore, the results received were compared with manual measurements done by two radiologists. The comparative analysis was carried out to examine the similarity of boundaries generated by the program and drawn by radiologists. Results depend on the accuracy and experience of a radiologist. The analysis (for noise free images) shows very similar and satisfactory results for both lower and upper boundary. The radiologists had the greatest problem with the extraction of the upper boundary. The comparative analysis (for noise free images, upper boundary) between radiologists shows that only about 40% of pixels lie in the area for d=4. The reason was the weak contrast.

The high similarity between radiologists and between program and radiologists were observed in the case of cluttered images for lower boundary. Unfortunately, the similarity between the upper boundaries drawn by radiologists and extracted by program was not satisfactory.

The estimation of maximums and inflection points is not proper for the cluttered images. In general, the estimation is not appropriate in places, where the undesired structures exist. They distort the true shape of the mandibular. The lower boundary extraction algorithm selects the pixels associated with the highest intensity region. This is sometimes a problem for cluttered images where the brightest region is not always associated with the next boundary pixel. The reason is the presence of the undesired structures (like the hyoid bone). In such situation, the algorithm can lose the appropriate path and extract an improper edge.

The cortical width measurements carried out by the program were satisfactory for the noise free images. The cortical width measurements were carried out often incorrectly for the cluttered images. The reason is the weak extraction of the upper boundary. The method is not robust enough to overcome the noise problem (the undesired structures presence).

9.9 SUMMARY OF CHAPTER

The method for measuring the cortical width on dental panoramic radiographs was presented in the present chapter. The technique does not require any training data as in [3]. Involvement of the operator (like a dentist) is less than in [4], [7].

The similarity of boundaries generated by the program and drawn by radiologists were satisfactory for the noise free images. The cortical width measurements computed by the program were satisfactory for the noise free images.

Both the maximum estimation and the inflection point estimation methods can work incorrectly if the cluttered areas are present in an image.

The presented method may be useful for screening patients with osteoporosis based in the case of noise free images.

9.10 EXERCISES

Exercise 1
What is the region of interest which is cut by an examiner (enumerate the features)?

Exercise 2
What is the cortical width? Show it on the image below.

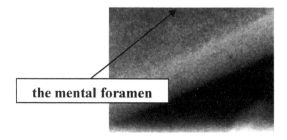

the mental foramen

Exercise 3
Extract the lower and upper boundary by hand. Draw the perpendicular line to the tangent to the lower cortical bone (the perpendicular line crosses the mental foramen).

Exercise 4
Implement the method of the gradient computing explained in the section 9.3.1. Compare the result with gradient images computed according to the equations given below

$$G(m,n) = \sqrt{|I(m+1,n) - I(m,n)|^2 + |I(m,n+1) - I(m,n)|^2}$$
$$G(m,n) = max(\|I(m+1,n) - I(m,n)\|, |I(m,n+1) - I(m,n)\|).$$

Use the following image.

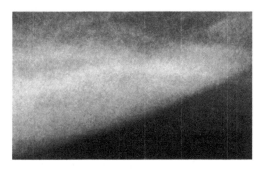

Exercise 5
Assume that the table temp contains the first maximums' locations, temp=[31 30 32 34 32 20 33 35 34 36 10 36 36 ...] (e.g. the second cell (number 30) means that the 30-th point of the second profile is the first maximum). The sequence is filtered using median filter (window span equals to 3). Determine the filtered sequence assuming that the nearest maximums in relation to the filtered sequence are [31 30 32 34 32 34 33 35 34 36 35 36 36 ...].

Exercise 6
What is the weights influence on the lower boundary extraction? Use weights below to check the results of the lower boundary extraction.

*- for left image side, **- for right image side.

Region	A	B	C	D	E	F
Weights*	1.0	2.0	2.0	2.0	2.0	1.0
Weights**	2.0	2.0	1.0	1.0	2.0	2.0

Exercise 7
Every k-th pixel is selected whereas the maximum estimation is carried out. Is the window span (median filtering) associated with the k value? Should the greater window span be used for greater k?

Exercise 8
Suppose that the set of the profiles is given. One profile is selected to measure the cortical width. Propose an algorithm of the appropriate profile selection based on pseudo-code. Assume that the mental foramen location is known.

Exercise 9
The picture below contains the undesired structure. The hyoid bone overlaps the mandibular bone. What are the difficulties of the upper boundary extraction?

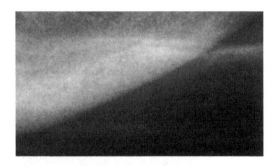

Exercise 10
How the comparative analysis is carried out? What is the Bland-Altman plot?

ANSWERS TO THE EXERCISES

Exercise1
The rectangular region of interest, which is cut by examiner with respect to the mental foramen includes part of mandibular structure as large part as possible to compute the cortical width. The area received in this way contains both the mandibular and the trabecular bone has smaller size than an original radiograph, which results in decreasing of the computation time.

Exercise 2
The cortical width is the Euclidian distance between the cortical bone boundary and the upper boundary (the bright area) and is determined between two points, which lie on the perpendicular line to the tangent to the lower cortical bone boundary as it is illustrated in Figure 9.2.

Exercise 3
See Figure 9.2.

Exercise 4
The result should be similar to the image in Figure 9.6 (step 1).

Exercise 5
Follow the explanation given in Section 9.5.1 p. 12 .
The sequence [...34 32 20 33 35 34 36 10 36 ...] filtered using median filter (window span equal 3) results in [...34 32 34 33 35 34 36 35 36 ...]

Exercise 6
The weights are associated with regions D, E, F to favor the directions of the search where the pixels of interest belonging to the lower boundary should be extracted. The region associated with the highest value is selected.

Exercise 7
The window span (median filtering) is not associated with k. Different masks can be used, the greater mask, the more attenuated noise.

Exercise 8
Follow the explanation given in Section 9.6 and Figure 9.12 .

Exercise 9
The curvature of the upper boundary is broken, therefore its appropriate extraction is very difficult and should be approved by a radiologist.

Exercise 10
A comparative analysis was done between the measurement of the cortical width by using the program and the measurement of the cortical width performed by the radiologists. The Bland-Altman plot is a graphical method to compare two measurement techniques. The x axis corresponds to the average of two measurements, while y axis corresponds to the difference between measurements. Horizontal lines are drawn at the mean difference, and at the limits of agreement, which are defined as the mean difference plus and minus 1.96 times the standard deviation of the differences

References

[1] M. Van der Klift , H.A. Pols, JM Gelenijse, D.A.M, Van der Kuip, A. Hofman, C.E.D.H De Laet, "Bone mineral density and mortality in elderly men and women," Bone, 30, pp. 643-651, 2002.

[2] K. Horner, H. Devlin, C.W. Alsop, I.M. Hodgkinson, J.E Adams, "Mandibular bone mineral density as a predictor of skeletal osteoporosis," British Journal of Radiology, 69, pp. 1019-1025, 1996.

[3] P. D. Allen, J. Graham, D. J. J. Farnell, E. J. Harrison, R. Jacobs, K. Nicopolou-Karayianni, C. Lindh, P. F. van der Stelt, K. Horner, H. Devlin, "Detecting Reduced Bone Mineral Density From Dental Radiographs Using Statistical Shape Models," IEEE Transaction on Information Technology in Biomedicine, 6, pp. 601-610, 2007

[4] A. Z. Arifin, A. Asano, A. Taguchi, T. Nakamoto, M. Ohtsuka, M. Tsuda, Y. Kudo, K. Tanimoto, "Computer-aided system for measuring the mandibular cortical width on dental panoramic radiographs in identifying postmenopausal women with low bone mineral density," Osteoporosis International, 17, pp. 753-759, 2006.

[5] K. Karayianni, K. Horner, A. Mitsea, L. Berkas, M. Mastoris, R. Jacobs, C. Lindh, P.F. van der Stelt, E. Harrison, J.E. Adams, S. Pavitt, H. Devlin, "Accuracy in osteoporosis diagnosis of a combination of mandibular cortical width measurement on dental panoramic radiographs and a clinical risk index

(OSIRIS): The OSTEODENT project," Bone, 40, pp. 223–229, 2007.

[6] E. N. Mortensen, W. A. Barrett, "Interactive Segmentation with Intelligent Scissors," Graphical Models and Image Processing, 60, pp. 349–384, 1998.

[7] A.Z. Arifin, , A. Asano, A. Taguchi, T. Nakamoto, M. Ohtsuka, K. Tanimoto, "Computer-aided system for measuring the mandibular cortical width on panoramic radiographs in osteoporosis diagnosis" , Progress in Biomedical Optics and Imaging - Proceedings of SPIE, 5747, pp. 813-821, 2005.

[8] http://developer.intel.com (Open Source Computer Vision Library, Reference Manual).

Chapter 10.

Development of a Computer Vision Application for Surgical Skill Training and Assessment

Gazi Islam[1], Kanav Kahol[1] and Baoxin Li[2]

[1]Department of Biomedical Informatics, Arizona State University, [2]Department of Computer Science, Arizona State University
Tempe, Arizona, USA
gislam@asu.edu; kanav@asu.edu; baoxin.li@asu.edu

ABSTRACT

Laparoscopic surgery requires many hours of systematic practice with a simulator to acquire psychomotor skills. Timely feedback on each simulated surgical session can facilitate surgical skills acquisition more efficiently and effectively than having no or delayed feedback. The current feedback system requires constant presence or post-hoc analysis on a recorded surgery of a competent surgeon or trainer to subjectively assess the surgical dexterity of the trainee by providing a composite score, which lacks inter-rater reliability. Several video and sensor-based systems have been developed to capture a user's motion and other important features which can be later analyzed objectively and quantitatively to correlate with the skill level. However, none of these systems is capable of providing real-time feedback, and they are all subject to several drawbacks. As a result, there is a need for a robust, real-time, quantitative surgical skill assessment system.

The proposed system uses a computer vision optical flow algorithm to analyze the surgical tool-video in real time, and provide dynamic feedback scores to assess surgical training performance on three key dimensions: motion smoothness, tool movement efficiency, and errors,. Since the feedback is

Bio-Informatic Systems, Processing and Applications, 191-206,

immediate, it is hypothesized that the proposed approach will increase a) training efficiency by reducing the training workflow cycle time (accomplished by shrinking the feedback response time) and b) training effectiveness by reducing the time it takes to accomplish a desired level of proficiency (improvements) on the three scores (smoothness, movement and error reduction). This technique can also be potentially useful to assess a surgeon's performance in an actual surgical operation.

10.1 INTRODUCTION

Due to the demand for greater accountability and patient safety in health care delivery, effective surgical performance measurement is gaining an increasingly high profile. Advances in the medical field with new methods, curricula, and processes, such as Accreditation Council for Graduate Medical Education competencies or Objective Structured Assessment of Technical Skills, as well as innovative technologies such as web-based learning and simulation have made surgical skill acquisition more challenging than ever [1]. To ensure the best surgical performance, systematic simulator training programs are being developed alongside traditional training in hospitals. It is a new and progressive way to intensify surgical resident training and surgical skills learning. The traditional training method needs constant presence of a competent surgeon to measure the progress of a trainee. This means of evaluating surgical dexterity is highly subjective and does not yield quantitative data [2]. Several studies have been done to address the issue of evaluating the user's performance in simulator-based training systems. In order to accurately measure a user's skill objectively and quantitatively, the system must satisfy the following requirements [3]: (1) the system must possess adequate sensing techniques to monitor the user's operation; (2) the system must extract relevant features from the sensing data; and (3) the system must have a good computational model to represent the skill demonstrated in the operation. Such a model is essential for accurately measuring the technical competence of the performance.

In the proposed research, a novel video-based approach is used to capture continuous, long sequences of a surgeon's hand, posture and surgical tool movements during an entire surgical procedure using inexpensive video camera. Video data is analyzed using a computer vision algorithm and then integrated to correlate with the user's skill level. For modeling the surgical skill, a stochastic approach is proposed. Finally a data mining technique is used to develop an observer-independent model through objective and quantitative measurement of surgical skills.

Computer vision has achieved many technological breakthroughs in the last few decades. It has been successfully used for object detection, tracking, motion detection and analysis, etc. As it is the focus of many opportunities to improve health care, it is exploring a wide array of new technologies. A variety of computer vision applications that have been invented in the past few years can be applied in clinical as well as other domains of biomedical informatics to solve many problems. In this research, an open source computer vision library (OpenCV) [4] has been used to track physicians' hand motion from captured video with the goal of assessing surgical dexterity from the analysis.

Tracking of hand, posture and surgical tool movement is one of the key characteristics used in evaluating surgical performance. Many sensor-based systems have been developed to accurately track a surgeon's hand or surgical tool movement for laparoscopic skill assessment [3, 5, 6]. However, all the present systems have several drawbacks. The focus of the proposed research is to develop a robust algorithm that is free from all of the following drawbacks:

- Sensor integration interferes with surgical skill execution [3].
  Most of the skill assessment systems available compelled surgeons to wear sensors to track different features of the surgical procedure that can be correlated with the skill execution. Since this system analyzes video data and does not require wearable sensors, it is free from any interference caused by sensors.
- Existing assessment techniques are system specific and so cannot be transferred to different simulators.
  Most of the skill assessment techniques available are designed for one particular simulator. Since they track different features in different exercises, each of these techniques is designed for one designated task only. However, the skill assessment system proposed in this research, evaluates a surgeon's skill by performing movement smoothness analysis. Smooth hand or surgical tool movement is required for successful execution of every surgical task, thereby making it a common skill assessment feature among most of the simulation tasks. Apart from detecting several task specific features, this research work mainly focuses on hand movement smoothness in surgical skill assessment which helps to develop a versatile algorithm that can be used in several surgical simulators.
- Since skill assessment can be performed only in simulations, the system may not be transferable to assess a real-life surgery.
  Surgeons perform surgery by looking at the video images which can be analyzed (even in real-time) to find movement smoothness, errors, and overall performance in skill execution. Video data of a surgeon's hand movement can be collected by mounting the video camera with the IV stand. Integration of sensor causes interference to skill execution. Moreover the sensors are expensive [7] and they can only increase the cost of the entire surgical procedure. Since there are no wearable sensors in the proposed system, it has no sterility issue.
- Video-based system has problem with lighting, background noise and poor image resolution [3].

The tool video data is collected from video screen where lighting is consistent inside the box simulator. In case of capturing hand movement, videos are normalized according to the size of the hand and lighting. By detecting individual objects, the video can be separated from background noise. Also, both the tool movement and hand movement are captured from a very short distance, hence ensuring high resolution video.

Efficient skill training in minimally invasive surgical (MIS) techniques has become standard in surgical residency programs. Unlike open surgery, MIS is, by nature, a technique that is very suitable for simulation-based training [8]. The specific psychomotor skills and eye–hand coordination needed for this type of surgery can be trained easily through box-trainers and computer-enhanced simulation trainers. The acquisition of laparoscopic skills requires a longer learning curve than does open surgery [9]. This research also focuses on making the learning curve steeper by providing on-screen real-time performance feedback to the user. The computer vision algorithm is able to detect some important features from the tool video such as motion analysis and error detection, and facilitates resident learning of psychomotor skills more efficiently (Figure 10.1).

10.2 BACKGROUND AND RELATED WORK

Minimally invasive surgery is now gaining acceptance as the main surgical procedure because of its known advantages over open surgery. The skills required for MIS are markedly different from those employed in open surgery. It is now generally accepted that these skills must initially be learned in training laboratories prior to entering the operating theater [10]. Mastery of laparoscopy skills is difficult. It requires many hours of practice and a certain amount of innate psychomotor skills. Major challenges in MIS are-

- Translation of the 2-dimentional image of the operating field from the video screen into a 3-dimentional mental image [11, 12].
- Learning to operate using long instruments.
- Mastering ambidexterity and eye-hand coordination.

Surgical technical progress in residency training is expected to improve with increased training and the repetition of certain procedures [8]. Transfer of training occurs when one learns a simple skill and adapts that skill to various complex skills [13]. Practicing with simple tasks like pegboard transfer or suturing help to acquire skills that can be transferred to actual surgery. Many commercially available simulators are being used in surgical residency program to teach MIS. Simulation-based surgical training can be categorized into three groups [14-16]:

- **Physical reality:** A traditional box simulator provides haptic feedback; however it requires constant presence of an expert surgeon to subjectively measure the progress of the trainee. The FLS Laparoscopic Trainer Box [17] is a widely used physical reality simulator for surgical residents and practicing surgeons that facilitates the development of psychomotor skills and dexterity required during the performance of basic laparoscopic surgery.
- **Augmented reality:** Augmented reality surgical simulators provide both realistic haptic feedback and objective assessment after the performance. ProMIS [18], computer-enhanced laparoscopic training system (CELTS), Blue Dragon [5, 6, 19] and LTS3e are some widely used augmented reality simulators. These augmented reality simulators use a variety of monitors, cameras, and complex computing equipment, which must be of high quality but also affordable.
- **Virtual reality:** VR simulators provide explanations of the tasks to be practiced and objective assessment of the performance; however they lack realistic haptic feedback. Muresan et al. showed in research that for novice trainees, the efficacy of VR training is questionable. In contrast, physical and augmented training methods had benefits in terms of time, quality, and perceived workload [20]

Most of the simulators use time as a metric for measurement of surgical proficiency and rewards faster performance with higher performance score [21]. Other measures of proficiency include kinematics of the laparoscopic tools. Tracking hand and surgical tool movement is one of the most important features used in assessing surgical performance and many sensor-based systems have been developed for this purpose. Pressure sensors were embedded in several simulators to observe the force applied by the surgeon while performing a surgical task [7, 22, 23]. A surgeon's respiration rate [24] was also examined in a study for any correlation useful to skill assessment. Some systems use generic measures such as smoothness of the movements as a measure of proficiency. The WKS system [22-24] measures force and movement of the dummy skin in a suture/ligature training system to evaluate performance. By using wireless sensor glove and body sensor network (BSN) technology [25], hand gesture data can be captured and analyzed with a Hidden Markov Model (HMM) for surgical skill assessment. Several systems have been developed to measure performance in actual surgery. The Wasada Bioinstrumentation (WB) system [26] uses a series of sensors to track head, arm and hand movement as well as several physiological parameters to analyze a surgeon's performance during laparoscopic surgery. Sadahiro et al. used a force platform to measure fluctuations of an operator's center of pressure (COP) [27] to estimate the skill level in the operating room. Most of these systems need multiple wearable sensors, which could interfere with an operator's skill execution. Also the sensors need to be sterile to be used in actual surgery and can make the entire surgery very expensive. Chen et al. proposed a video-based system to track special markers on the glove [3]. However it requires consistent lighting and direct line-of-sight to the markers which might not be possible during actual surgery.

Different algorithms are used to assess surgical skill from all these data. All these algorithms can be classified into 3 major categories- regression models, data mining algorithms and statistical analyses. Regression models use data as parameters and develop equations to provide a scoring mechanism for the surgical tasks [7, 22, 23, 27-30]. When multiple dimensional data are observed, various data mining tools are used to reduce the dimensionality in order to observe the clusters in a two or three dimensional plane to differentiate among the skill levels [3, 5, 6, 19, 25, 31-34]. This is because when observing data with relatively lesser dimensionality, calculating simple statistical mean and variance could be sufficient to detect significant differences in skill levels [2, 24, 26, 35].

Despite its obvious importance in surgical teaching programs, the assessment of surgical skills has been studied infrequently and inadequately. Munz et al. performed objective evaluation of the performance on a series of simulated tasks such as open abdominal, laparoscopic, and intubation skills between a small group of junior and senior trainees. The results have shown no significant differences between participants for all procedural tasks regardless of grouping, level of training, or total number of years in training [36]. Feldman et al. compared objective assessment of technical skills and subjective in-training evaluations of surgical residents. However the experimental result was statistically insignificant [37].

10.3 METHODOLOGY

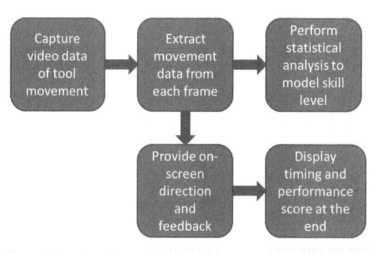

Figure 10.1 Methodology.

The OpenCV program is used to extract surgical tool movement data from the video of a simulated surgical procedure. The program uses the combination of object detection and motion segmentation algorithms to find how an image changes over time. The first algorithm uses predetermined threshold values to detect specific objects (Figure 10.2). Surgical tool video from box trainer is captured and by using the object detection algorithm each of the objects is separately detected (Figure 10.3).

The second algorithm uses motion history images from the silhouette of the object to find the gradient of the motion. It detects any subtle movement and leaves a trail of the motion (Figure 10.4).

The stationary part of the video remains dark (Figure 10.5) and by counting the number of pixels in each frame, the velocity of tool movement is measured. An array of acceleration of hand movement is then created for each user, further differentiating this value. This gross measurement analysis is called the analysis of movement smoothness which is one of the most important features in any surgical procedure.

255	255	255	255	255	255	255	20
255	255	255	100	100	255	20	20
255	255	255	100	100	255	20	20
255	255	255	100	100	255	20	20
255	255	255	255	255	255	20	20
255	255	255	255	255	255	255	255
150	150	255	255	255	255	255	255
150	150	255	255	255	255	255	255

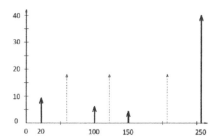

Figure 10.2 Histogram of gray-scale image.

Figure 10.3 Object Detection in Surgical exercise.

Figure 10.4 Motion Detection.

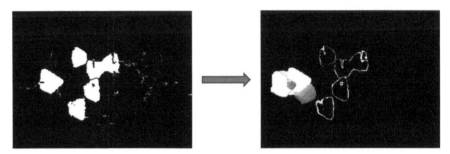

Figure 10.5 Motion capture from video.

Optical flow also shows the gradient of resultant movement (Figure 10.6). The program counts the number of times the tools have accelerated to complete the task, finds the degree of each acceleration value and prepares an array for it. This array measures the economy of tool movement which is another important feature in surgical skill assessment.

Figure 10.6 Gradient of movement.

Both of these raw movement values are analyzed in real-time and used for providing on-screen movement feedback. The program also displays the time elapsed since the beginning of the surgical procedure and records the completion time for each task. Since OpenCV is already used for detecting different objects in the tool video, it is also used to monitor the progress of the surgical task, provide direction to the user and detect the completion of each unit of task (Figure 10.7).

The program automatically tracks the tasks according to the predetermined order and detects any error performed during the process (for example: dropping an object or picking a wrong object), counting it towards the composite score. Finally it shows a result page with a summary of the performance (Figure 10.8).

10.4 RESULTS

Participants from Mayo Clinic, Rochester, MN of 3 different post-graduate years performed peg transfer, intracorporeal suturing and knot tying using an FLS box [17]. Hand tremor value for individual participants was extracted using computer vision algorithm and then normalized. The value of redundant hand motion was also derived from the hand tremor values and is presented in the bar-charts below.

 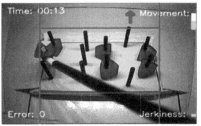

Figure 10.7 Simulator Training without feedback (left) and with on-screen feedback (right).

Total Time: 01:03
Total Drop: 1
Movement: 1.1%
Jerkiness: 0.9%

Figure 10.8 Result page after completion of the task.

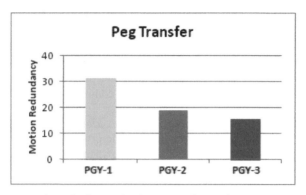

Figure 10.9 Average motion redundancy value for peg transfer exercise.

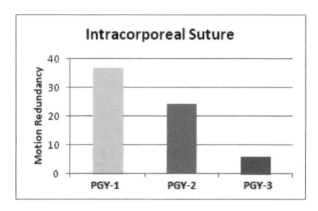

Figure 10.10 Average motion redundancy value for intracorporeal suture exercise.

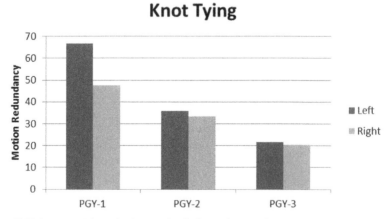

Figure 10.11 Average motion redundancy value for knot tying exercise.

For all three surgical tasks, a significant difference in hand motion redundancy was observed [38] (Table 10.1). Motion redundancy is shown as the percentage of absolute motion in the following bar-charts (Figures 10.9-10.11).

A data set from Mount Sinai Hospital, New York, of a simulated intubation exercise has 14 users- 3 attending surgeons, 3 surgical residents (2 PGY-2s and 1 PGY-1) and 8 medical students (Table 10.2). The entire dataset was analyzed and the average motion redundancy value for each group of users is shown in Figure 10.12. The amount of hand acceleration significantly increases as the expertise level goes from attending to surgical resident to medical student. Moreover it is noticeable that the acceleration value is higher for the "difficult" level than the "easy" level of the same task [40].

Table 10.1
T-test analysis of Mayo clinic data

t-test	p-value
PGY-1 : PGY-2	0.1427
PGY-2 : PGY-3	0.0492
PGY-1 : PGY-3	0.0297

Table 10.2
Mount Sinai participants

Intubation	
Level	No of participants
Medical Students	8
Surgical Residents	3
Attending Surgeons	3

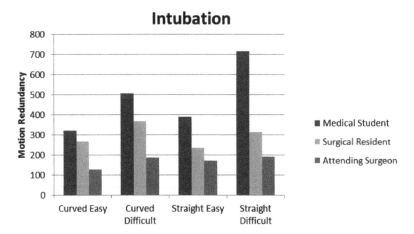

Figure 10.12 Average acceleration value of user groups.

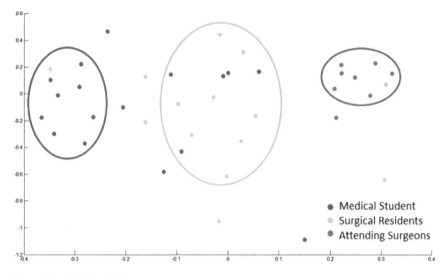

Figure 10.13 LDA Analysis.

The LDA analysis of the acceleration data is shown in Figure 10.13. User's data roughly conform to 3 clusters. One of the surgical residents fall into expert's cluster whereas another falls into novice's cluster.

T-test analysis of the 3 user groups was also performed using Matlab. The t-test returned a highly significant p-value (**p = 0.0032**) between attending and medical students (Table 10.3). Since the performances of resident fall into both expert and novice cluster, the p-value is comparatively higher (thus less significant) than for the other 2 groups.

10.5 EXPECTED OUTCOME, LIMITATIONS AND CHALLENGES

Video data without the implementation of a feedback system has been already collected and analyzed. New data with the on-screen feedback system is being collected from the Simet Center (Banner Good Samaritan) and Mayo Clinic Hospital, Phoenix, AZ. Video-based skill assessment algorithm will be tested with both the datasets. Data will be tested for significant improvement in learning curve of acquiring psychomotor skill over the previous approach without on-

Table 10.3
T-test analysis of Mount Sinai data

t-test	p-value
Attending : Med Student	0.0032
Attending : Resident	0.0202
Resident : Med Student	0.0705

screen feedback. Data has also been collected from Arizona State University (ASU) graduate students to compare the learning curve when the feedback system is introduced to completely novice subjects. The outcome of this research should provide a robust algorithm for assessing surgical skill and strong evidence of efficient skill learning with the proposed on-screen real-time feedback.

The main limitations and challenges of the video-based system are the following:

- Resolution: High resolution video is preferred in extracting information from video. However resolution is sometimes compromised for distance. Getting a close distance video recording might be challenging in case of real surgery.
- Lighting: Difference in lighting can change the video quality substantially and thus create difference in analysis of data. Shading from the tools has adverse effect on the video analysis too.
- Background noise: The video captures any movement in the background with the foreground or hand/tool movement. It causes noise while analyzing the video. To minimize this noise, surgeons were asked to wear particular colored gloves and later by selecting proper threshold in the histogram, only the gloved hand movements were extracted. The use of a tool such as Microsoft Kinect can minimize background noise as well. Since it carries depth information, background can be easily distinguished from foreground.

10.6 CONCLUSION

This research proposes a video-based surgical skill assessment technique. It uses a computer vision algorithm to analyze the video of hand and surgical tool movement and extract features like hand-movement smoothness and task specific errors. Finally, it uses a supervised learning technique for the classification of skill level. Data from different PGY level of surgical residents and expert surgeons is being collected to train and test the algorithm. Since the research uses video data rather than any wearable sensors, it is cost effective and also overcomes the drawbacks of most surgical skill assessment techniques presently available.

Since the proposed video-based surgical skill assessment technique can provide real-time on-screen feedback, it is also being tested as a tool for the efficient skill learning technique. After analyzing data from the experiment with on-screen feedback system, the result shows steeper learning curve than the system without the feedback. However more data is being collected for analysis to strengthen the hypothesis.

Therefore, this is a cost-effective and robust surgical skill assessment technique which also promotes toward efficient surgical skill learning. However, the key for successful execution of skill assessment is to deploy the algorithms in a real surgical setting. Until now almost no work has been done to transfer this work into an actual surgical setting. Apart from induced interference in surgical execution, the sensor-based systems needs to be completely sterile to be used in actual surgery. Also the extensive use of sensors makes the entire surgery and the training procedure more expensive. Since the video-based approach is free from the drawbacks of the other systems, this technique can be potentially useful to assess a surgeon's performance in an actual surgical operation. Real-life surgical videos are collected to perform some testing of the system in actual surgery. However, further analysis and research need to be done in this area.

References

[1] Satava, R., The Revolution in Medical Education - The Role of Simulation. Journal of Graduate Medical Education 2009. 1(2): p. 172-175.
[2] Dosis, A., ct al. ROVIMAS: a software package for assessing surgical skills using the da Vinci telemanipulator system. in Information Technology Applications in Biomedicine, 2003. 4th International IEEE EMBS Special Topic Conference on. 2003.
[3] Chen, J.Y., M., Sharma, R. , Visual modelling and evaluation of surgical skill. Pattern Anal Applic, 2003. 6: p. 1-11.
[4] Bradski, G. and A. Kaehler, Learning OpenCV: Computer Vision with the OpenCV Library2008, Sebastopol, CA: O"Reilly Media, Inc.
[5] Rosen, J., et al., Generalized approach for modeling minimally invasive surgery as a stochastic process using a discrete Markov model. Biomedical Engineering, IEEE Transactions on, 2006. 53(3): p. 399-413.
[6] Rosen, J., et al., Markov modeling of minimally invasive surgery based on tool/tissue interaction and force/torque signatures for evaluating surgical skills. Biomedical Engineering, IEEE Transactions on, 2001. 48(5): p. 579-591.
[7] Aizuddin, M., et al. Development of Sensor System for Effective Evaluation of Surgical Skill. in Biomedical Robotics and Biomechatronics, 2006. BioRob 2006. The First IEEE/RAS-EMBS International Conference on. 2006.
[8] Derossis, A.M., et al., The effect of practice on performance in a laparoscopic simulator. Surgical Endoscopy, 1998. 12(9): p. 1117-1120.
[9] Peter D. Vlaovic, M., et al., Immediate Impact of an Intensive One-Week Laparoscopy Training Program on Laparoscopic Skills Among Postgraduate Urologists. Jounals of the Society of Laparoendoscopic Surgeons, 2008. 12(1): p. 1-8.
[10] Aggarwal, R., et al., A competency-based virtual reality training curriculum for the acquisition of laparoscopic psychomotor skill. The American Journal of Surgery, 2006. 191(1): p. 128-133.
[11] Pamela B. Andreatta, E., et al., Laparoscopic Skills Are Improved With LapMentor™ Training - Results of a Randomized, Double-Blinded Study.

Annals of Surgery, 2006. 243(6): p. 854-863.

[12] Rabiya Suleman, M., et al., Hand-Eye Dominance and Depth Perception Effects in Performance on a Basic Laparoscopic Skills Set. Jounals of the Society of Laparoendoscopic Surgeons, 2010. 14(1): p. 35-40.

[13] Figert, P.L., et al., Transfer of training in acquiring laparoscopic skills. Journal of the American College of Surgeons, 2001. 193(5): p. 533-537.

[14] Botden, S. and J. Jakimowicz, What is going on in augmented reality simulation in laparoscopic surgery? Surgical Endoscopy, 2009. 23(8): p. 1693-1700.

[15] Choy, I. and A. Okrainec, Simulation in surgery: perfecting the practice. The Surgical clinics of North America, 2010. 90(3): p. 457-73.

[16] Jakimowicz, J. and A. Fingerhu, Simulation in surgery. British Journal of Surgery, 2009. 96(6): p. 563-564.

[17] Fundamentals of Laparoscopic Surgery...the definitive laparoscopic skills enhancement and assessment module, http://www.flsprogram.org/.

[18] The world-leading ProMIS Surgical Simulator.

[19] Rosen, J., et al. The BlueDRAGON - a system for measuring the kinematics and dynamics of minimally invasive surgical tools in-vivo. in Robotics and Automation, 2002. Proceedings. ICRA '02. IEEE International Conference on. 2002.

[20] Muresan Iii, C., et al., Transfer of training in the development of intracorporeal suturing skill in medical student novices: a prospective randomized trial. The American Journal of Surgery, 2010. 200(4): p. 537-541.

[21] Derossis Md, A.M., et al., Development of a Model for Training and Evaluation of Laparoscopic Skills. The American Journal of Surgery, 1998. 175(6): p. 482-487.

[22] Solis, J., et al. Development of a sensor system towards the acquisition of quantitative information of the training progress of surgical skills. in Biomedical Robotics and Biomechatronics, 2008. BioRob 2008. 2nd IEEE RAS & EMBS International Conference on. 2008.

[23] Solis, J., et al. Quantitative assessment of the surgical training methods with the suture/ligature training system WKS-2RII. in Robotics and Automation, 2009. ICRA '09. IEEE International Conference on. 2009.

[24] Oshima, N., et al. Acquisition of quantitative data for the detailed analysis of the suture/ligature tasks with the WKS-2R. in Information Technology Applications in Biomedicine, 2007. ITAB 2007. 6th International Special Topic Conference on. 2007.

[25] King, R.C., et al., Development of a Wireless Sensor Glove for Surgical Skills Assessment. Information Technology in Biomedicine, IEEE Transactions on, 2009. 13(5): p. 673-679.

[26] Zecca, M., et al. Using the Waseda Bioinstrumentation System WB-1R to analyze Surgeon’s performance during laparoscopy - towards the development of a global performance index. in Intelligent Robots and Systems, 2007. IROS 2007. IEEE/RSJ International Conference on. 2007.

[27] Sadahiro, T., et al. Laparoscopic skill measurement with COP to realize a HAM Scrub Nurse Robot system. in Systems, Man and Cybernetics, 2007. ISIC. IEEE International Conference on. 2007.

[28] Kahol, K., M. Smith, and J. Ferrara. Quantitative Benchmarking of Technical Skill Sets as a Credentialing Strategy. in American College of Surgeons 96th Annual Clinical Congress. 2010. Washington, DC, USA.

[29] Lacey, G., et al., Mixed-Reality Simulation of Minimally Invasive Surgeries. Multimedia, IEEE, 2007. 14(4): p. 76-87.

[30] Witzke, W.O., et al. A criterion-referenced approach to assessing perioperative skills in a VR environment. in Advanced Learning Technologies, 2001. Proceedings. IEEE International Conference on. 2001.

[31] Kahol, K. and M. Vankipuram. Hand motion expertise analysis using dynamic hierarchical activity modeling and isomap. in Pattern Recognition, 2008. ICPR 2008. 19th International Conference on. 2008.

[32] Mackel, T.R., J. Rosen, and C.M. Pugh, Markov Model Assessment of Subjects' Clinical Skill Using the E-Pelvis Physical Simulator. Biomedical Engineering, IEEE Transactions on, 2007. 54(12): p. 2133-2141.

[33] Megali, G., et al., Modelling and Evaluation of Surgical Performance Using Hidden Markov Models. Biomedical Engineering, IEEE Transactions on, 2006. 53(10): p. 1911-1919.

[34] Reiley, C.E., E. Plaku, and G.D. Hager. Motion generation of robotic surgical tasks: Learning from expert demonstrations in Engineering in Medicine and Biology Society (EMBC), Annual International Conference of the IEEE. 2010.

[35] Wolpert, S., et al. Assessing motion in laparoscopic tools. in Bioengineering Conference, 2001. Proceedings of the IEEE 27th Annual Northeast. 2001.

[36] Munz, Y., et al., Ceiling effect in technical skills of surgical residents. The American Journal of Surgery, 2004. 188(3): p. 294-300.

[37] Feldman, L.S., et al., Relationship between objective assessment of technical skills and subjective in-training evaluations in surgical residents. Journal of the American College of Surgeons, 2004. 198(1): p. 105-110.

[38] Islam, G. and K. Kahol, Application of Computer Vision Algorithm in Surgical Skill Assessment, in IEEE 6th International Conference on Broadband Communications & Biomedical Applications (IB2COM) 2011: Melbourne, Australia.

[39] Islam, G. and K. Kahol, Development of Computer Vision Algorithm for Surgical Skill Assessment, in The 2nd International ICST Conference on Ambient Media and Systems2011: Porto, Portugal.

[40] Islam, G., K. Kahol, and e. al, Understanding specialized human motion through gesture analysis: Comparisons of expert and novice anesthesiologists performing direct laryngoscopy using motion capture technology in International Anesthesia Research Society Meeting2011: Vancouver, Canada.

Chapter 11.

Information Delivery System for Deaf People during Large Disasters

Atsushi Ito[1], Hitomi Murakami[2], Yu Watanabe[3], Masahiro Fujii[3], Takao Yabe[4] and Yuko Hiramatsu[5]

[1]*KDDI Research and Development Laboratories, Japan*
[2] *Seikei University, Japan*
[3]*Utsunomiya University, Japan*
[4]*Tokyo Metropolitan Hiroo Hospital, Japan*
[5]*Chuo University, Japan*
ito@kddilabs.jp, hi-murakami@st.seikei.ac.jp, yu@is.utsunomiya-u.ac.jp,
fujii@is.utsunomiya-u.ac.jp, takao_yabe@tmhp.jp,
susana y@tamacc.chuo-u.ac.jp

11.1   INTRODUCTION

5:46 am, January 17th, 1995 — a large earthquake at a magnitude of 7.3 occurred near Kobe and caused a catastrophic disaster.   Cities near the center of the earthquake were shaken with a seismic intensity of 7.   6,434 lives were lost and 43,792 were injured. 274,181 houses/buildings were broken; 7,500 were burned. 350,000 people took refuge in shelters; water supply was stopped for 130,000,000 families: electric power was stopped for 2,600,000 families: gas supply was stopped for 860,000 families and 300,000 telephone services were stopped.

We made a survey of individuals with auditory handicaps requiring support after the Great Hanshin–Awaji earthquake [1] and received 185 replies.   We found many of the deaf were left without support during the disaster. Some of them were left in a house and could not go to shelter. Even in a shelter, life was not comfortable, the battery was dead, the equipment was broken and the surrounding was very noisy.   They also required battery for their hearing aids, large displays for information, radios and mobile phones. Lack of these gadgets meant that it was difficult for them to get important information. 58.1% of the deaf were informed of the disaster immediately: however, 49.1% knew of the disaster with a delay of half a day. 3.7% knew about the disaster after a week. 44% of the persons who answered the questionnaires were those who had taken refuge in shelters; however, 30% of those answered that their hearing aid did not work, while 20% could not understand what the officials in the shelter said. We also did a similar survey after the Western Tottori earthquake [2] and had a similar result.

Of course, blind persons also had trouble and difficulties in events of disaster. It is, however, easy for us to distinguish the blind from others, since they usually hold a white cane. On the other hand, it is not easy to notice the deaf in a shelter, since they appear normal.

In the event of a disaster, usually electric power supply would be stopped. Deaf people normally secure information from captions of TV programs, faxes and PCs; however, these devices do not work when power fails. When power fails, transistor radio and portable loudspeaker are used to provide information.

On the other hand, ICT gears are very popular in Japan. Especially, about 78% of the population of Japan accesses the Internet (Figure 11.1) [3]. Also, the number of mobile phone users is 1.16 million, that is, calculated simply, 91% of the population (Figure 11.2) [3]. It is essential to use ICT equipment and to increase literacy and to prepare for catastrophic disaster to help the disabled.

Based on these survey results, we developed "Information Delivery System for Deaf People in a Larger Disaster (IDDD)" using mobile phones and ad hoc network technology. We performed several trials [8,9,10] and got good results for commercial release. IDDD displays disaster information on both mobile phone and display units that are designed for offices and homes. A message arriving at a mobile phone is transferred automatically to a display in a room by using Bluetooth. LED displays hanging on large walls are used in an office or public space, and small box type displays are used in residences.

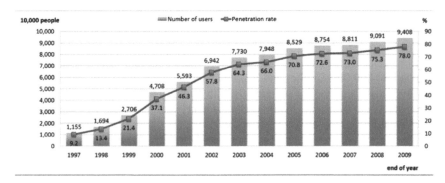

Figure 11.1 Changes in the number of Internet users and the penetration rate [3].

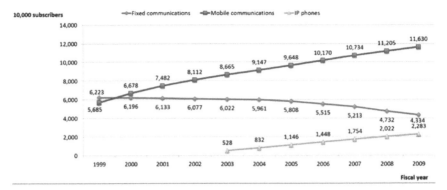

Figure 11.2 Changes in the number of subscriptions to fixed communications and mobile communications [3].

First of all, we will explain related system in Section 11.2, then we will discuss about the system requirements and how to implement them briefly in Section 11.3. In Section 11.4, we will mainly explain the result of the trial done at the beginning of 2010 and will wrap up this research done in the past three years.

11.2 RELATED WORKS

Not many researches relating to disaster information systems for the disabled or elderly persons have been performed. In this section, we will introduce some previous researches and working systems and will discuss how IDDD intends to solve the problems of existing systems.

SEMA4A ontology and relating researches [4,5,6] present interesting ideas. They proposed ontology to reduce victims of accidents and diseases. Ontology presents a system of accessibility to information in case of accident for various people such as the disabled, elderly persons or tourists under different circumstances such as fire or traffic accident. Also, they discussed various devices that can be used in case of accidents and diseases.

Taking as basis the SEMA4A ontology, they developed a prototype for automatically creating and sending personalized emergency notifications using different media and devices.

Emergency notification is transferred by combination of email, SMS, MMS, fax, phone etc. Also, they proposed effective combination of media for various kinds of profile of people such as deaf, blind, color blind, elderly, dementia, etc.

Information is displayed by character, picture, sound, vibration, movie etc.

This research provides excellent ontology not only for deaf but also for various other kinds of disabilities. However, they are not focusing on specialty of large disaster. We are focusing on large disasters that cause failure of electric power, broken lifeline, less information received by humans.

Table 11.1
Disaster Information System in Japan

System name	Problems
EyeDragonII	Does not work without TV Only works in living room
Information delivery to mobile phone by email (Kobe Disaster Prevention Net)	During dinner, taking bath, it is difficult to see mobile phone
IT-based Disaster Prevention System (JEITA)	Not focus on disabled
Wireless IP Phone (Owase City)	Not focus on disabled

In Japan, there are several disaster information systems for deaf people as described in Table 11.1. However, they have some problems. For example, EyeDragonII is located in living room, but it does not work when electricity fails, since it is related to TV set. Also, some disaster prevention systems do not solve the problems of deaf people. So, we developed IDDD to solve problems of deaf people in case of disaster. In the next section, we will discuss the requirements in detail.

11.3 SYSTEM REQUIREMENTS AND DESIGN

The main theme of this research is displaying disaster information more legibly. In other words, information should be displayed reflecting situation.

Disaster information system like IDDD should be used daily for every person to know that a useful display is there and waiting for him to help. Also IDDD-like system should work with a mobile phone since 84% of the Japanese population own mobile phones. We must make the most of the high penetration of mobile phones for providing disaster information.

According to our research on IDDD, one of the most important issues is to secure the daily use of disaster information display. To attain the target, we designed our IDDD, based on the following:

(1) Disaster information
(2) Daily information (Time, Weather, News etc.)
(3) Office information

As described in Table 11.2, there are several parameters for display such as communication methods, features of display such as color, velocity and quality of sound.

Table 11.2
Disaster information and display method

Information	Communication Method	Display		
		Color	Velocity	Sound
Disaster	Broadcast IP SMS Local	RED Flash	Fast	Alarm
Office	IP, SMS Local	GREEN	Slow	Comfortable
Daily Life (Time, Weather, News etc.)	IP	GREEN	Still or Slow	

Table 11.3
Features that are required for disaster information system and Media that can be used

#1	Media	IT vs Human (Voice / Paper)
#2	Power	Battery (including solar power) vs AC
#3	Distance	Mobile vs LAN vs Local communication (Bluetooth etc.)
#4	Immediately	IT vs human

Next, we discuss features of communication system that are indispensable in disaster situation and the method to satisfy these requirements as described in Table 11.3. ICT or mobile phone is not the perfect answer. For example, when we need to send information immediately, in some cases it is better to shout to attract attention rather than finding a PC, then to turn it on and finally to send email..

It is important that the most appropriate method should be selected during a disaster. Our IDDD is not a perfect tool, but we can use it in several ways to meet the situation. We would like to discuss it further in the following sentences.

Figure 11.3 explains the differences in media (#1 in Table 11.3), especially sound and characters. Our system mainly uses characters to provide information to deaf people. Also, information in characters is useful for normal people, in case of the failure to catch audio information in confusing situation.

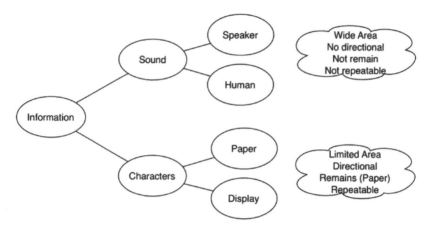

Figure 11.3 Classification of media for disaster information.

Electric power is of great importance for ICT devices (#2 in Table 11.3). In disaster situation, power supply is lost usually, so many ICT devices such as LAN and PCs should be stopped. We decided to use this system with battery if electric power supply is failed. Usually, mobile phone and mobile PC can work by battery and LED display of IDDD can do the same, though for a period of time. PCs, LAN and Display should work by AC power supply. Combination of mobile phone and display with battery is the only solution to meet disaster situation as described in Figure 11.4.

Also, combination of mobile network and fixed network realizes four different paths to access display to provide information as described in Figure 11.5 by using two different WAN (mobile and fixed) and two different local communication (#3 in Table 11.3). In addition, displays are connected each other as ad hoc network using Bluetooth [7].

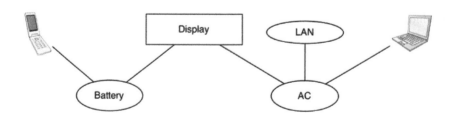

Figure 11.4 Robustness of network in case of power failure.

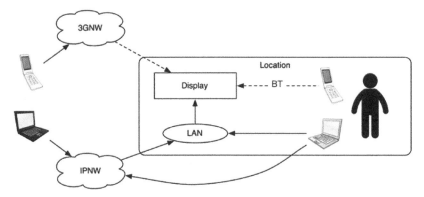

Figure 11.5 Multiple Accesses.

As described in #4 in Table 11.3, we should consider how to immediately transfer information. There are many possibilities to transfer information to people in deserted places. We would like to discuss three typical possibilities: one is transferring information by voice or paper (display), second is through LAN or ad hoc networking and the last is through WAN as described in Figure 11.6. In the disaster situation, the surest way is using voice or paper, since there is no need to switch on devices (PCs), applications and type in messages. The second fastest way might be to send messages by using LAN or ad hoc networking, since there is no network delay and ad hoc networking with mobile phone can provide strong tool.

Finally, we decided to implement the following issues in IDDD.

(1) Display should be visible and consume less power (#1, #4 in Table 11.3)
(2) Display should be carried by hand (#3 in Table 11.3)
(3) Display should work by battery (#2 in Table 11.3)
(4) Display should be controlled by mobile phone network (#3 in Table 11.3)
(5) Display should be connected using local wireless such as Bluetooth (#3, #4 in Table 11.3)

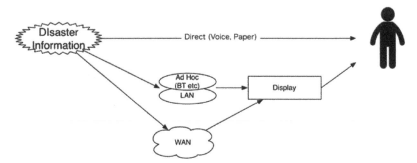

Figure 11.6 Immediate transfer of information.

In addition, we decided to install LAN connection on the display to display time or news. Since the display is set to display disaster information, there is very rare chance to use display. And people forget that there is a display to help them. Therefore, it is very important to setup a display of IDDD as a useful device in daily life.

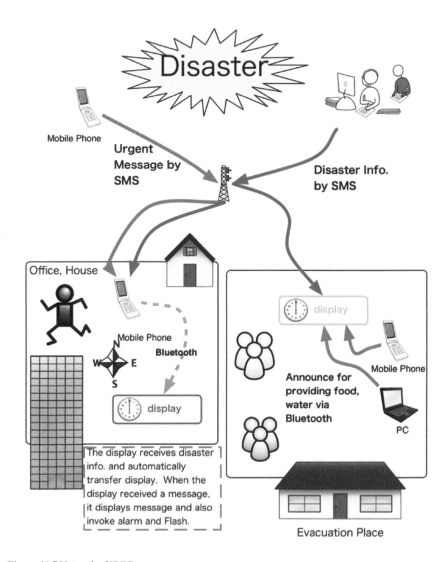

Figure 11.7 Network of IDDD.

Figure 11.7 explains the whole network of IDDD. We use SMS to send information since on the mobile handset, SMS can invoke application to relay received information to display.

11.4 EVALUATION

The main theme of this research is to display disaster information more legibly. In other words, information should be displayed reflecting situation.

11.4.1   Outline of evaluation

We have developed and evaluated IDDD from 2008 to 2010 under the support from "Association of Technical Aid". In 2008, we evaluated its usage in disaster events. We presented large size displays at several exhibitions or events and asked visitors to our booth to fill in questionnaires, to investigate the needs and the reactions to the response time and legibility of the display. In 2009, we developed a small box type display and a large display as in 2008 for IDDD. We tested the large one in the office and small one in the living room of the residence. Under the collaboration with Tokyo Metropolitan Center for Welfare of the Physically Handicapped, we performed a month long experiment to evaluate the usage in daily life. In 2010, we tested our system for a month as a call message board for the reception of otorhinolaryngology clinic of National Rehabilitation Center for Persons with Disabilities to evaluate the usage of IDDD in a public space. In this section, we describe the outline of experiment in 2010.

11.4.2   Description of the experiment

In Tables 11.4 and 11.5, stakeholder of this experiment and function of IDDD used in this experiment are described, respectively.

Figures 11.8 and 11.9 explain the actual display that was used in the experiment of 2010. Since the location was a hospital with busy doctors, they wanted the next patient to enter the consultation room when consultation of previous patient is finished. So we introduced a simple model, where the doctor inputted the token number of the next patient on his mobile phone. Then the mobile phone sent that number to the display via Bluetooth instead of inputting the number through PC by using application in Figure 11.10. This application was implemented using BREW [11].

In the experiment, we set two criteria as follows.

(1) Effectiveness in disaster for both deaf people and normal people
(2) Effectiveness in daily use for both deaf people and normal people

Table 11.4
Stakeholders of experiment

Information Provider	Emergency case: Training for providing emergency information such as fire or earthquake	KDDI R&D Laboratories
	Usual case: (1)Display the number of next patient (2)General announce	Doctor of otorhinolaryngology clinic
User	Hearing disabled and his/her escort	Patients
System Manager	Set up and operation	KDDI R&D Laboratories

Table 11.5
Functions of IDDD in this experiment

Mobile phone for Information Provider	Emergency case: Function to provide emergency information
	Usual case: Input two digits number to call next patient Input short message in 30 Japanese characters
Display	Display message from information providers by Bluetooth. In addition, there are following functions. Flash was installed on the left and the right side of the display to signal the arrival of new message Buzzer sound also worked as the signal for the new message
Server	Provide a long message of disaster (longer than 30 characters) This server was located in KDDI R&D Laboratories.

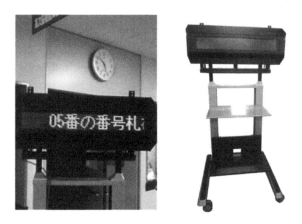

Figure 11.8 Display that was used at the experiment of 2010.

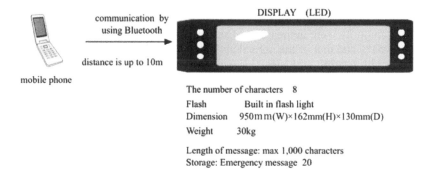

Figure 11.9 System configuration for the experiment of 2010.

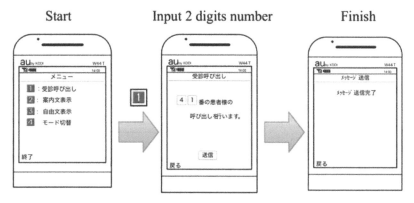

Figure 11.10 Application on mobile phone to input the number that is linked to next patient and send it to the display.

11.4.3   Messages

The display has two modes for message: one is emergency mode (RED) and the other is normal mode (YELLOW).   We set the following messages for experiment.

(Emergency) "This is training: A big earthquake happened."

(Normal-1) "No.XX, Please enter the room" (see Figure 11.8)

(Normal-2) "Office is closed today"

(Normal-3) "It is crowded, please wait here."

11.4.4   Results

1) System operation - There was no trouble during the experiment.
2) Results from the questionnaire
   a) Demography of participants
      The main participant of this trial was a young mother having a child with hearing problems.   70% of the participants were females and the majority were in their 30s. The ratio of those with hearing problems was 51% and that of the normal at 49%. Normal persons accompanied those with hearing problems. Also, elderly over 60 years of age account about 36%. The details of demography are shown in Table 11.6.

Table 11.6
Demographic of users

Total		46
Sex	Male	14
	Female	32
Age	10~19	4
	20~29	2
	30~39	16
	40~49	2
	50~59	5
	60~69	8
	70~79	7
	80~89	1
Hearing	Deafness	1
	Difficulty in Hearing	8
	Slight Difficulty in Hearing	14
	No Problem	22
	NA	1
with hearing aid	Yes	16
	No	30

b) Evaluation

We have tested IDDD for three years since 2008 that showed IDDD had been improved. Figure 11.11 presents the overall evaluation of this system in 2010. Forty six users, that is 87%, agreed on the usefulness; 80% of the users agreed on the font size; 89% of the users agreed on the brightness; 96% of the users agreed on the colors of the character; 76% of the users agreed on the scroll speed; and 87% of the users agreed on the readability. So, we can say that this system received better result, compared to the results of 2008 and 2009. Figure 11.12 shows the result of 2009. It shows that effectiveness was also good, but several requests to improve font size and brightness were made. In 2010, as described in Figure 11.11, brightness and font size are improved and 86% of the users expressed positive evaluation as a whole.

In 2009, one big problem was regarding the brightness of the display. We changed the display from the flat vacuum fluorescent tube display to the LED display with anti-reflecting film. A buzzer was installed to signal the arrival of new information to normal persons. Thanks to these improvements, the evaluation of the effectiveness of this system was very high as described in Figure 11.13: "Very Useful + Useful" was 87% and the evaluation of readability of the display was also high, "excellent, good" being 72%.
Some said that melodious chime would be better than noisy buzzer.

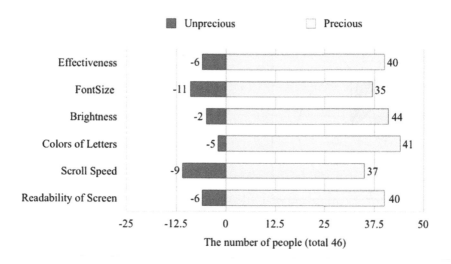

Figure 11.11 Over all evaluation of experiment in 2010.

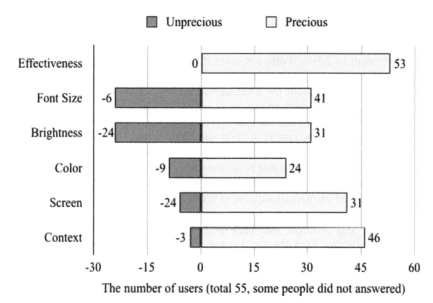

Figure 11.12 Over all evaluation of experiment in 2009.

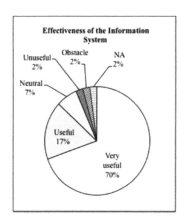

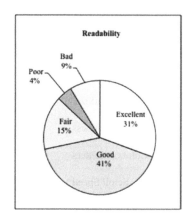

Figure 11.13 Effectiveness of the system and Readability of the display.

11.5 CONCLUSION

In this chapter, we discussed how our information delivery system for deaf persons in disaster (IDDD) has been designed and improved. The following criteria was taken up for IDDD system.

- Display should be visible and consume lesser power
- Display should be carried by hand
- Display should work by battery
- Display should be controlled by mobile phone Network Display should be connected using local wireless such as Bluetooth

In particular, brightness and font size are improved and 86% of user agreed that this system is useful. Also, "Very Useful + Useful" was 87% and the evaluation of readability of the display was also high, "excellent, good" being 72%.

It was expected that the development of this IDDD would be completed and launched as a commercial product in 2010.

Our next step, however, is to extend information delivery system to a network of people as described in Section 11.3, the power of human being is an integral aspect in cases of disasters. It is also true that human power and network are required in daily life. The IDDD is designed to help deaf or elderly people by using the power of ICT in local society.

Recent models of mobile phones have several new functions to help elderly and handicapped persons, such as mail reader and screen reader. Also, digital signage has become popular. We think that systems like IDDD should be useful, if they are installed in many places in a town such as show window of shops. It is usually used for advertisements, however, it could be used as a sign board to direct the nearest shelters.

In disaster, mobile phones and public signboards should work together to overcome the emergency situation. We would like to extend our research to such direction.

On March 11th, 2011, a large earthquake and Tsunami hit Tohoku area of Japan. After the disaster, we started a trial to use IDDD at Deaf Association Miyagi and a school for the deaf in Sendai City as preparation for aftershock. Also, we are now extending IDDD to use Android phone [12]. Earlier implementation was based on BREW, so that the mobile phones that could be used were limited. Instead, by using Android, we can use IDDD application on various kinds of phones. Also Android allows us to collaborate with other applications easily by using "internet".

References

[1] "Survey of individuals with auditory handicaps requiring support after the

Great Hanshin-Awaji earthquake", T.Yabe, Y.Haraguchi, Y.Tomoyasu, H.Henmi, A.Ito, Japanses Journal of Disease Medicine, Vol.14, No.1, 2009

[2] "Survey of individuals with auditory handicaps requiring support after the Western Tottori earthquake", T.Yabe, Y.Haraguchi, Y.Tomoyasu, H.Henmi, A.Ito, Japanses Journal of Disease Medicine, Vol.12, No.2, 2007

[3] Information and Communications in Japan, Ministry of Internal Affairs and Communications, Japan, 2009,
http://www.soumu.go.jp/johotsusintokei/whitepaper/eng/WP2010/2010-index.html

[4] "SEMA4A: An ontology for emergency notification systems accessibility", A.Malizia, T.Onoratia,P.Diaza,I.Aedoa,and F.Astorga-Palizaa, Expert Systems with Applications, Volume 37, Issue 4, April 2010, Pages 3380-3391

[5] "CAP-ONES: An Emergency Notification System for all", Alessio Malizia, Pablo Acuña, Teresa Onorati, Paloma Díaz, Ignacio Aedo, Proceedings of the 6th International ISCRAM Conference – Gothenburg, Sweden, May 2009

[6] "Emergency Alerts for all: an ontology based approach to improve accessibility in emergency alerting systems", Alessio Malizia,Francisco Astorga-Paliza,Teresa Onorati,Paloma Díaz,Ignacio Aedo, Proceedings of the 5th International ISCRAM Conference – Washington, DC, USA, May 2008

[7] "Mobile phone based ad hoc network using built in Bluetooth for ubiquitous life", Hitomi Murakami, Atsushi Ito, Yu Watanabe, Takao Yabe, Proc. The 8th International Symposium on Autonomous Decentralized Systems (ISADS 2007), March 2007

[8] "An Information Delivery and Display System for Deaf People in Times of Disaster", Atsushi Ito, Hitomi Murakami, Yu Watanabe, Masahiro Fujii, Takao Yabe,Yoshikura Haraguchi, Yozo Tomoyasu, Yoshiaki Kakuda, Tomoyuki Ohta, Yuko Hiramatsu, Proc. Telhealth2007 (May 2007)

[9] "A study on deaf people supporting systems using cellular phone with Bluetooth in disasters", Masahiro Fujii1, Amir Khosravi Mandana, Takatoshi Takakai, Yu Watanabe, Kazuo Kmata, Atsushi Ito, Hitomi Murakami, Takao Yabe, Yoshikura Haraguchi, Yozo Tomoyasu, Yoshiaki Kakuda, Proc. EXPONWIRELESS (June 2007)

[10] "Universal Use of Information Delivery and Display System using Ad hoc Network for Deaf People in Times of Disaster", Atsushi Ito, Hitomi Murakami, Yu Watanabe, Masahiro Fujii, Takao Yabe, Yoshikura Haraguchi, Yozo Tomoyasu, Yoshiaki Kakuda, Tomoyuki Ohta, Yuko Hiramatsu, Proceedings of Broadbandcom2008, pp.486-491

[11] "brewmp", http://www.brewmp.com/

[12] "Android", http://www.android.com/

Acronyms

ADL	Activity of Daily Living
API	Application Programming Interface
ASU	Arizona State University
ATP	Automatic Train Protection
BSN	body sensor network
CELTS	computer-enhanced laparoscopic training system
COP	center of pressure
DHCP	Dynamic Host Configuration Protocol
DWT	Discrete Wavelet Transform
ECG	electrocardiogram
EEG	Electroencephalography
EMS	electromyographic signal
FAID	Fatigue Audit Inter-Dyne
FFD	Full Function Device
FFT	Fast Fourier Transform
GCC	GNU Compiler Collection
GCK	Gaussian distribution of Clustered Knowledge
GPRS	General Packet Radio Service
HDMI	High Definition Multimedia Interface
HF	high frequency
HMM	Hidden Markov Model
HSDPA	High Speed Data Packet Access
IAV	Integrated Absolute Value
ICA	Independent Component Analysis
ICT	Information and Communications Technology
IDDD	Information Delivery System for Deaf People in a Larger Disaster
IMNF	Instantaneous Mean-Frequency
IP	Internet Protocol
KNN	k-nearest neighborhood
LDA	Linear Discriminant Analysis
LF	low frequency
LoG	Laplacian of Gaussian
MAV	Mean Absolute Value
MCT	Means Comparison Test
MDF	median frequency

MES	myoelectric signals
MIS	minimally invasive surgical
MLP	Multilayer Perceptron
MNF	mean frequency
MNS	mean scale
MUAP	motor unit action potentials
NCA	Neighborhood Component Analysis
NSW	New South Wales
NTSC	National Television System Committee
OpenCV	open source computer vision library
PAL	Phase Alternating Line
PC	principal component
PCA	principal component analysis
PS2	PlayStation-2
PS3	The PlayStation 3
RFD	Reduced Function Device
RMS	Root Mean Square
SCAL	scalogram
sEMG	Surface electromyography
SI	Separability Index
SMS	Short Message Service
SSC	Slope-Sign Changes
SSL	Secure Sockets Layer
STFT	Short Time Fourier Transform
TDOA	Time Difference of Arrival
TOR	Trained Observer Rating
TV	television
UniBw	Universität der Bundeswehr München
USB	Universal Serial Bus
VAR	Variance
WAMP	Willison Amplitude
WB	Wasada Bioinstrumentation
WCC	worst-case crossover
WFL	Waveform Length
WHO	World Health Organization
WL	Waveform Length
WSNs	Wireless Sensor Networks

ZC	Zero Crossings
ZC	ZigBee Coordinator
ZED	ZigBee End Device
ZR	ZigBee Router

INDEX

A

Activity of Daily Living, 73
alert state, 128, 129, 130, 149
alpha, 127, 128, 129, 130, 131, 132, 133,
     134, 142, 143, 147, 148, 150, 151,
     152, 153, 154, 155, 156, 157, 158,
     160, 161, 162, 163, 164, 167
Android, 221, 222
Association of Technical Aid, 215
auditory handicaps, 207, 221, 222

B

battery, 28, 87, 207, 212, 213, 221
beta, 127, 128, 129, 130, 131, 132, 133,
     134, 142, 143, 162, 163
biomedical, 2, 53, 54, 55, 65, 66, 193
Bluetooth, 61, 208, 211, 212, 213, 215,
     216, 221, 222
BMD, 170, 171
breast, 1, 5, 6, 20, 21, 22, 23, 24
breast cancer, 1, 5, 6, 20, 23
BREW, 215, 221
buzzer, 219

C

cancer, 1, 2, 3, 4, 5, 6, 8, 20, 21, 22, 23,
     24, 25
Canny, 173
class separability, 108, 116, 117, 119,
     121, 122
Clustering, 73
Coordinator, 86, 89

D

Daily information, 210
Data Analysis, 50
Data Clustering, 73
deaf, 207, 208, 209, 210, 211, 215, 221,
     222

delta, 128, 134, 142, 143

diagnosis, 1, 2, 4, 5, 6, 7, 8, 9, 20, 21,
     22, 23, 24, 25, 170, 171, 189, 190
digital signage, 221
Dimensionality Reduction, 51, 52, 107,
     121
disaster, 207, 208, 209, 210, 211, 212,
     213, 214, 215, 216, 221
Disaster information, 210, 211
Disaster Prevention System, 210
Discrete Wavelet Transform, 72, 73
display, 57, 162, 164, 208, 210, 211,
     212, 213, 214, 215, 216, 217, 218,
     219, 220, 221
Drowsiness, 139, 149, 160, 161, 162

E

earthquake, 207, 216, 218, 221, 222
ECG, 56, 142, 146, 160
EEG, 127, 128, 129, 134, 135, 136, 139,
     142, 143, 146, 147, 148, 150, 151,
     152, 153, 154, 156, 158, 159, 160,
     161, 162
emergency mode, 218
Emergency notification, 209
ensemble neural network, 80
equation $(\theta+\alpha)/\beta$, 127, 130, 134, 135

F

Fall Detection, 69, 70, 80
Fast Fourier Transform, 128, 151, 162
fatigue, 30, 31, 33, 34, 41, 42, 43, 44,
     45, 46, 47, 49, 54, 65, 66, 112, 125,
     126, 127, 128, 129, 130, 134, 135,
     136, 139, 140, 142, 144, 159, 160,
     161
fatigue countermeasure, 127, 134, 135,
     160
Feature Extraction, 73
Feature Selection, 51
Fourier transform, 29, 110, 147

G

gait phase, 109, 121
Gait Phase, 123
game consoles, 55, 59, 66

Gaussian, 69, 71, 75, 80, 112, 173, 179
Gaussian Distribution, 69, 75, 80

About The Editors

Johnson Agbinya holds PhD from La Trobe University (1994) in microwave radar remote sensing (MSc Research University of Strathclyde, Glasgow, Scotland (1982) in microprocessor techniques in digital control systems) and BSc (Electronic/Electrical Engineering, Obafemi Awolowo University, Ife, Nigeria (1977). He is currently Associate Professor (remote sensing systems engineering) in the department of electronic engineering, Honorary Professor in Computer Science at the University of Witwatersrand in Johannesburg, South Africa and Extraordinary Professor in telecommunications at Tshwane University of Technology/French South African Technical Institute in Pretoria, South Africa. He was Senior Research Scientist at CSIRO Telecommunications and Industrial Physics (1994 – 2000; renamed CSIRO ICT) in biometrics and remote sensing and Principal Engineer Research at Vodafone Australia research from 2000 to 2003. He is the author of seven technical books in electronic communications including Principles of Inductive Near Field Communications for Internet of Things (River Publishers, Aalborg Postkontor, Denmark, 2011); IP Communications and Services for NGN (Auerbach Publications, Taylor & Francis Group, USA, 2010) and Planning and Optimisation of 3G and 4G Wireless Networks (River Publishers, Aalborg Postkontor, Denmark, 2009).

He is Consulting Editor for River Publishers Denmark on emerging areas in Telecommunications and Science and also the founder and editor-in-chief of the African Journal of ICT (AJICT) and founder of the International Conference on Broadband Communications and Biomedical Applications (IB2COM). His current research interests include smart environments, remote and short range sensing, inductive communications and wireless power transfer, Machine to Machine communications (M2), Internet of Things, wireless and mobile communications and biometric systems.

Dr. Agbinya is a member of IEEE, ACS and African Institute of Mathematics (AIMS). He has published extensively on broadband wireless communications, inductive communications, biometrics, vehicular networks, video and speech compression and coding, contributing to the development of voice over IP, intelligent multimedia sub-system and design and optimisation of 3G networks. He was recipient of research and best paper awards and has held several advisory roles including the Nigerian National ICT Policy initiative.

Edhem Custovic received a B.E. Hons. (Telecommunications) degree from the School Of Engineering & Mathematical Sciences, La Trobe University, Melbourne, and a Masters of Entrepreneurship and Innovation at Swinburne University, Melbourne, in 2007 and 2011 respectively. He is currently undertaking a Ph.D (Electronic Engineering/Space Physics) at La Trobe University under Prof. John Devlin and is a research associate in the Department of Electronic Engineering. Edhem was an electrical engineering consultant at FMG Engineering (2008-2010) and is currently an engineering project manager with Engineering First. Edhem has secured numerous research/non-research grants; from the AutoCRC and Generations (Sustainability Department of La Trobe University). His research interests lie in HF & VHF Radars, Remote Sensing, Antenna theory/design and Sustainability Engineering. He is part of the TIGER (Tasman International Geospace Environment Radars) research team in the Department of Electronic Engineering at La Trobe University and has also supervised numerous honours and masters engineering project students. Edhem was the founder of the La Trobe University IEEE student branch and was the inaugural Chair. He currently holds the position of Vice-Chair of the IEEE Victorian Section and was the founder of the IEEE tutorial program. He is also heavily involved in student mentoring programs and is a graduate of La Trobe University's Infinity Leadership Program. He has authored over 20 peer reviewed publications and is regular reviewer for several international conferences. His research and technical contribution has played a key role in the development of state of the art digital HF radar, TIGER-3.

Jim Whittington is a Senior Lecturer in Electronic Engineering at La Trobe University, Melbourne, Australia. He graduated with BSc(hons) and MEng degrees in 1984 and 1996 respectively.

Jim is currently working on his PhD entitled "Digital TIGER – A Digital HF Radar for Improved Sea State and Space Weather Monitoring". Jim is a member of the TIGER radar group that operates two SuperDARN class HF ionospheric radars in Tasmania and New Zealand. He is the design manger of the electronic systems for the next generation "all digital" TIGER radar, due for deployment near Adelaide early in 2011. Jim's research interests cover the application of Electronic Design Automation tools and techniques to communication, HF radar and control applications which particularly include real-time hardware implementation of Digital Signal Processing structures in Field Programmable Gate Array technologies.

Associate professor Sara Lal holds Science degree from the University of Sydney, Masters in Applied Science (University of Technology Sydney) (UTS)), PhD (UTS), Diploma in Law (Legal Profession Admission Board, Sydney) and the Graduate Certificate in Higher Education (UTS) from Australia. A/Prof Sara Lal is a neuroscientist and academic in the School of Medical and Molecular Biosciences at UTS with expertise and research areas spanning drowsiness and sleep, cognitive function, medical and neurophysiology and investigating medical biomarkers and countermeasures. A/Prof Lal supervises the research team and students in the Neuroscience Research Unit at UTS.

A/Prof Lal is a co-author of two books including multiple book chapters in Advances in Broadband Communications and Networks (River Publishers Series in Communications, River Publishers, Aalborg Postkontor, Denmark, 2008) and Biomedical and Environmental Sensing (River Publishers Series in Information and Science Technology, River Publishers, Aalborg Postkontor, Denmark, 2009). A/Prof Lal collaborates with various medical and IT industries on research and has won multiple competitive grants previously and has published journal and conference papers in areas of driver drowsiness and technology, medical biomarkers and physiology.

Lightning Source UK Ltd.
Milton Keynes UK
UKOW06n1936141214

243130UK00002B/48/P